W9-CLA-964

Test of FAITH

Test of FAITH

Science and Christianity Unpacked

Study Guide

Edited by Ruth Bancewicz

WIPF & STOCK · Eugene, Oregon

www.testoffaith.com

TEST OF FAITH: Study Guide
Science and Christianity Unpacked

Wipf & Stock
An imprint of Wipf and Stock Publishers
199 W. 8th Ave., Suite 3
Eugene, OR 97401
www.wipfandstock.com

ISBN: 978-1-60899-896-8

Cover design by Contrapositive
Print Management by Adare

Contents

Introduction

This study guide is designed to accompany the "Test of Faith" course. It will help you to follow the discussion, take notes if you like, and will provide some background reading and follow-up materials.

The challenge that has been put forward so many times recently is that God is a delusion and science has removed the need for faith in anything. But there are many practicing scientists who have a sincere Christian faith, even at the highest levels of academia. They have all been trained to think and test ideas to the limit. If their faith and their science are both genuine searches for truth, we need to hear from them. All of us can share their experience of awe when they find out more about God's creation through science. These scientists also help us to explore a number of issues that affect our own lives.

The book ***Test of Faith: Spiritual Journeys with Scientists*** tells the stories of ten of the scientists interviewed for the documentary – their personal discovery of faith and how that has affected their work.

Christians hold different views on some of these issues. We're not expecting everyone to agree with all that the scientists and theologians say in the documentary. These are complex issues, and in the course material we have often laid out several different views that Christians take on a particular topic so that you can discuss them openly.

This material is very much an introduction to science and Christianity. It may well provoke some questions that you didn't know that you had and open up some new areas of interest. The course includes many different questions and points for discussion. They cannot *all* be covered in each session, but there are plenty of good books, articles and mp3s available to follow up topics that you did not cover in the course sessions or take things further, some of which are recommended in the "Taking it Further" lists for each session. There are also extra resources tailored specifically for the course on **www.testoffaith.com**.

Session 1: Beyond Reason?
Science, Faith and the Universe

SESSION 1

Preparation for Participants *(optional):*
Why bother thinking about science and faith?

Ask two or three friends, family members or colleagues if they can think of a situation where science and religion (or beliefs) affect each other. What issues or questions arise?

For example, what about:
- In medicine? (Religious beliefs often affect ethical decisions.)
- In education? (Children sometimes ask questions like "Who made human beings, God or evolution?")
- In politics? How does a candidate's views on environmental issues affect your vote?

or

When you watch TV, listen to the radio, or read the paper, keep an eye out for stories that mention both science and faith. What issues or questions arise? What effect do these issues have on society?

Glossary	
atheist	Someone who believes that no gods exist.
cosmologist	Someone who scientifically studies the origin, development and overall shape and nature of the universe.
fine-tuning (the Anthropic Principle)	The idea that the physical constants of the universe are set at the precise values necessary for the existence of biological life.
God of the gaps	An argument which says that when we can't explain something in nature scientifically, that is proof that God exists.
metaphysic	Any particular way of interpreting the world.
multiverse theory	The idea that there are multiple universes. Some people use this to argue that if there are many universes, it's not so surprising that one of them is "fine-tuned" for life.

Discussion Topics

? How Do People See the World?

Dr. Ard Louis said that the debate between science and religion is really about how we decide whether something is true or false: is science the only reliable way of finding things out about the world, or does religion have something valuable to contribute as well?

Q1: With this in mind, what views have you heard from scientists (either in the media or that you have met personally) on the questions, "Does God exist?" and "How does God interact with the universe?"

There's a difference between scientific evidence and the interpretation of that evidence. It is possible for people of any religion or none to come up with the same results when they run the same experiment – but how do you interpret that evidence? Obviously some interpretations will be more reliable than others. The truth of a particular interpretation can be tested with more experiments.

Q2: Which of the views above (scientism/atheistic materialism, deism or Christianity) do you think could fit with the evidence the scientists have described about the universe in the documentary?

The Big Bang

In this chapter of the DVD the narrator asks the question, "Hasn't the Big Bang done away with the need for a Creator?"

Q1: What do you think? Do you think God could have created through the Big Bang?

SESSION 1

God of the gaps?

> " If you say, well, science answers this much about the way the universe is, but science doesn't answer this aspect of the universe's characteristics, and then to invoke God and to **allow God to reside in that gap in our knowledge, that's dangerous** because when a clever scientist comes along that gap will be filled by a deeper and richer scientific understanding. So then, where you posit that God is allowed to reside, gets smaller and smaller and smaller, and this is a practice known as **God of the gaps,** and it's dangerous.
>
> Professor Katherine Blundell

Read Colossians 1:15–17 and Hebrews 1:3a.

Q2: Where do we see God at work? What do these passages say about this idea of not putting "God in the gaps"?

Prayer and Miracles

"God of the gaps" raises the question: "Do scientists, even if they are Christians, believe that God cannot, or will not, work miracles?" Not at all. Here is why:

> " While some biblical miracles, such as … the plague of locusts in Egypt, do not directly contradict the laws of nature, other miracles are obviously supernatural. So does science challenge our belief in miracles? … **Christians believe that *God*, not natural laws, governs nature.** God typically works through natural laws to sustain the regular patterns of our world, but nature is not locked into those patterns.
>
> Deborah and Loren Haarsma, *Origins* (Faith Alive Christian Resources, 2007), p. 41.

> " **I believe God can choose to step out of his regular pattern and do something different at times,** but it would be for a reason relating to an answer to prayer, or something about God's desire to interact spiritually with his people. So I see miracles of healing, miracles in human history, or miracles in the Bible that God used to establish his chosen people and develop a relationship with them. It seems less likely that God would do miracles in natural history that we couldn't discover until modern science.
>
> Deborah Haarsma, *Test of Faith: Spiritual Journeys with Scientists*, p. 97.

Q3: How would you define a "miracle"?

Q4: How does this match your own experience or knowledge of the Bible?

But are these scientists filling the gap with science instead of God? Is this "science of the gaps"?

There are always unanswered questions in science. Christians working in science are aware that there could be a "supernatural" explanation for something being the way it is, but the early scientists – who believed in a Creator God – believed that they should investigate creation, and they found answers to their questions. Rather than mourning the apparent loss of mystery in creation, they rejoiced that they understood God's creation a little bit more and were able to praise God for creating the details they had just uncovered. Christians who are scientists still follow in that tradition, and it is their job to keep looking for answers to their further questions. The more we find out, the further we realize we have to go.

Q5: How do you think these two concepts, miracles and "God of the gaps," fit together for a Christian?

SESSION 1

Fine-tuning

Q1: What do you think of the idea of fine-tuning? Do you think it's reasonable to say that the universe as we know it is finely tuned for life (the Anthropic Principle)?

Q2: Earlier, the documentary warned us not to believe in a "God of the gaps." But could the idea of fine-tuning be a gap? Think about the difference between the evidence for fine-tuning and the gaps where people have put God in the past (e.g., as an explanation for changes in the weather).

Professor Alister McGrath thinks that fine-tuning is consistent with the idea of a God but that it doesn't prove God's existence.

Q3: Even if fine-tuning *were* evidence for the existence of God, why couldn't it completely prove God's existence?

Q4: If fine-tuning was evidence for the existence of God, what kind of God would it be evidence for?

How Do You View Science and Faith?

Some of the most common ways of relating science and faith are:

1. They're in conflict: they ask the same questions and get different results.
2. They're the same: *either* faith can be explained entirely by science, or science can be explained by faith.
3. They're complementary: science and faith ask different questions, and there are some areas where they can overlap and interact in positive ways.
4. They're non-overlapping: they ask different questions.

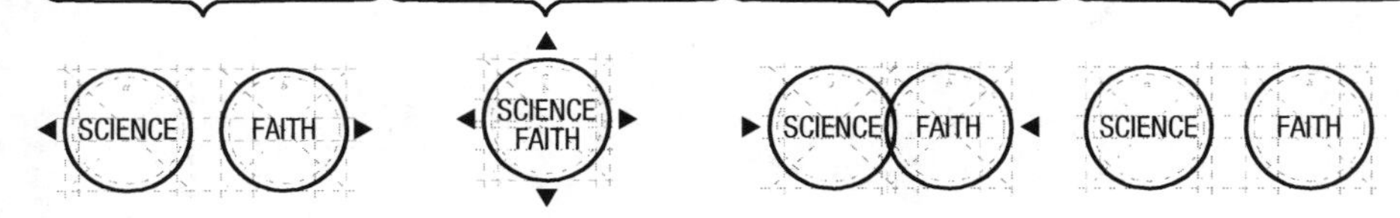

Q: Which do you most agree with, and why?

If you can think of a different way of relating science and faith, sketch your own diagram below.

SESSION 1

Taking it Further

Websites *(general introductions)*:

Talks, short papers and links: **www.faraday-institute.org**

Articles, interviews and help for Christians working in or studying science (UK): **www.cis.org.uk**

Articles, interviews and help for Christians working in or studying science: **www.asa3.org**

Articles to download on the topic of the universe:

John Polkinghorne, "The Anthropic Principle and the Science and Religion Debate": **www.st-edmunds.cam.ac.uk/faraday/resources/Faraday%20 Papers/Faraday%20Paper%204%20Polkinghorne_EN.pdf**

Rodney Holder, "Is the Universe Designed?": **www.st-edmunds.cam.ac.uk/faraday/resources/Faraday%20 Papers/Faraday%20Paper%2010%20Holder_EN.pdf**

Michael Poole, "God and the Big Bang":
http://www.cis.org.uk/assets/files/articles/Poole_bang.pdf

Rodney Holder, "God, the Multiverse, and Everything":
http://www.cis.org.uk/assets/files/Resources/Articles/Article-Archive/rodney_holder_multiverse.pdf

Books *(general introductions):*

Denis Alexander and Robert S. White, *Science, Faith, and Ethics: Grid or Gridlock?* (Hendrickson, 2006)
This book introduces the issues – including ethics, the environment and evolution.

Kirsten Birkett, *Unnatural Enemies* (Matthias Media, 1997)
This introduction to the relationship between science and faith is written for people with no Christian background.

Francis Collins, *The Language of God: A Scientist Presents Evidence for Belief* (Free Press, 2006)
This is a very easy introduction and includes much of the former director of the Human Genome Project's personal journey from atheism to Christianity. Covers evolution, ethics and the general relationship between science and faith.

Books on the universe:

David Wilkinson, *God, Time and Stephen Hawking* (Monarch, 2001)
This is an easy but thought-provoking introduction to science and theology.

John Polkinghorne, *Quarks, Chaos and Christianity* (Crossroad, 2005)
This is John Polkinghorne's most introductory level book. Much of it focuses on physics and astronomy.

Bonus Session 1b: Beyond Reason?
Facts and Faith

Preparation for Participants *(optional):*
Why bother thinking about science and faith?

It would be worth doing the preparation for Session 1, if you did not use it already, or:

As a recap of some of the issues covered in the last session, and as an introduction to this session, read one of the articles recommended on the "Taking it Further" list for Session 1 ("God and the Big Bang" is the most introductory of these).

Discussion Topics

Facts and Faith

What would you say to someone who said, "Surely science is about rationality and faith is about irrationality"?

Watch the bonus interview: 1.1 Ard Louis and John Polkinghorne

Glossary	
fideistic	Something that is based solely on faith or revelation, ignoring reason or intellect.

Q1: Can you think of a decision in your life that had an almost entirely rational or logical basis?

Q2: Can you think of a decision that involved faith, or trust?

Read Luke 1:1–4 and 1 Corinthians 15:1–11.

Q3: As Ard Louis and John Polkinghorne say in the DVD, logic and faith often overlap. Can you talk about one significant step or decision in your own life and how rationality or faith influenced that decision?

Intelligibility

Watch the bonus interview: 1.2 Ard Louis

Ard Louis refers to the "deep logos or logic behind the universe."

Read John 1:1–3. The "word' in John 1 is a translation of the Greek word *logos,* from which we get the English word "logic." It's generally agreed that John took the Greek idea of *logos*, a mind or rational principle governing the universe, and said that *logos* is Jesus, who was there in the beginning and came to earth as man.

Q1: What do you think of the idea that we can do mathematics because a rational God created the world?

A Personal God

Read John 1: 4–5.

Watch the bonus interview: 1.3 John Polkinghorne and Katherine Blundell

Q2: All of the scientists interviewed have an experience of a personal God. What evidence or experiences show this personal God is: A) for Christians in general; and B) for you?

The Big Bang

Watch the bonus interview: 1.4 Deborah Haarsma

Q1: A light year is the distance light travels in a year (about 6 trillion miles). Imagine light leaving the earth when you were born. What was happening in the world then? If you imagine TV pictures of those things being beamed out across the universe, where will they have reached now?

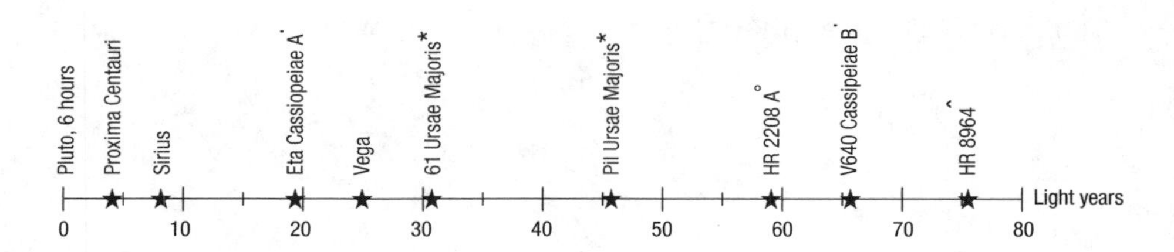

* In the Great Bear constellation

° In the constellation Orion

· In the constellation Cassiopeia

^ In the constellation Andromeda

You can think about this the other way around, too. If you use a telescope to look at the night sky, some of the light that you can see began its journey when you were born. Around which star is that light coming from?

 **Watch the bonus interview: 1.5 David Wilkinson**

Glossary	
cosmology	The study of the origin, development and overall shape and nature of the universe.
order of magnitude	Most commonly used to mean ten times larger (e.g., 5,000 is two orders of magnitude larger than 50).

How have Christians responded to Big Bang theory? Here are three views that Christians hold:

A. God made the universe and everything in it supernaturally about 10,000 years ago, as described in Genesis 1. The scientific evidence for the Big Bang and the great age of the universe is faulty. Creation scientists have suggested other possible explanations for the vast scale of the universe and the distance light must travel.

B. God made the universe and everything in it supernaturally about 10,000 years ago, as described in Genesis 1. Scientists are correct about the evidence for the age of the universe, but this is only an appearance of history. Everything was created with this history built in: trees with rings representing hundreds of years, light already on the way from stars billions of light years away, etc.

C. The scientific evidence for the Big Bang is correct. God used the Big Bang to make the universe 13.7 billion years ago and has been sustaining it ever since. Christians holding this view have different ways of reconciling this with Genesis 1, which will be discussed in Session 2.

Q2: Which of these views have you come across before?

The Multiverse

 **Watch the bonus interview: 1.6 Deborah Haarsma and David Wilkinson**

Glossary	
Particle physics	The study of the tiny particles that make up atoms.

Q1: What are some of the different scientific responses to the idea of the multiverse that the documentary gives?

Q2: Several of the interviewees mention how they feel about the idea of the multiverse as Christians. What are their views? Based on your knowledge of the Bible, do you think there are any arguments for or against the idea of the multiverse?

SESSION 1

 **Watch the bonus interview: 1.7 David Wilkinson**

Q3: What is the difference between evidence and proof? Can you think of any examples?

Q4: Keeping all this in mind, what do you think is the most helpful way to look at evidence in science and relate that to faith?

Taking it Further

See list for Session 1.

Session 2: An Accident in the Making?

Creation, Evolution and Interpreting Genesis

Introduction to Session 2

Genesis addresses questions that we all ask: "Why are we here?" "What are we here for?" "What makes us special?" There has been much debate about how to interpret Genesis, and there are important questions here for everyone to consider about the origin of human life, the presence of evil in the world and suffering.

Part 2 of the documentary looks briefly at several views on creation before moving on to examine critically the way that some Christians reconcile evolution with the Bible. It then moves beyond the debate to consider one live issue that all Christians have been called to tackle: caring for creation.

Note: The aim of the documentary is to show what the majority of Christians working in science believe, and to examine those beliefs in detail. We felt that it was important to mention other views, but we have not covered these in equal depth because they have already been well communicated and debated in print, film and online. Instead, the documentary examines very closely the implications of "Theistic Evolution."

The fact that God created the universe is one of the vital messages of Christianity. God is capable of acting in any way he chooses. God is intimately involved with his creation – not only by sustaining everything, but also by making direct and personal contact with the people within it, including "supernatural" miracles when he chooses (see Session 1). The resurrection is the most important of these miracles. While all Christians are united in their belief in God as Creator, believing in any particular *mechanism* of creation is not a benchmark of orthodox Christianity and none of the great church creeds specify this. As such, it is an important subject, open to careful, informed debate, and one about which Christians can hold different views without threatening their unity. The course will equip you to discuss a range of views on creation and come to your own conclusions.

This is not a detailed Bible study on Genesis. Its aim, rather, is to help people to digest the information that the documentary presents and to explore the biblical material underlying the points that it raises. For a more detailed study of Genesis, there are a large number of study guides and commentaries available.

Session 2: Overview

Preparation for Participants *(optional)*: Views on creation

If you had to explain your own views on creation, what would you say?

or

Ask two or three friends, family members or colleagues what they think about how the universe, with our planet and all the life we see on it, came to be.

Glossary	
DNA	The chemical molecule inside every cell of every living thing that carries the instructions for its growth and development.
Intelligent Design	The idea that some parts of living things are too complex to have evolved, coupled with the idea that the information contained in DNA cannot have arisen by a process describable in purely material terms, so providing evidence for "design."
Theistic Evolution (or Evolutionary Creationism)	The belief that God created life through the process of evolution.
Young Earth Creationism	The belief that Genesis should be interpreted as a literal, historical and scientific account, and therefore that God created the world between 6,000 and 10,000 years ago in six twenty-four hour days.

Discussion Topics

Views on Creation

Read Genesis 1.

Q1: What are the most important points in this passage? What does it say about God? About God's relationship with the universe? About God's relationship with people?

Q2: What do you think of the different views on creation that the documentary presented? Were any of them new to you?

Q3: Thinking back to the main points you picked out in Genesis 1, how would you say the views on creation that you just discussed in Q2 complement or contradict the biblical material?

Human Evolution

Read Genesis 2:4–25 and the Briefing Sheet: Views on Genesis 2 and 3, Who were Adam and Eve?

Q1: Which of these views do you think could fit best with the biblical and scientific account?

So how are we special compared to other animals? The Bible says that we are made "in God's image." We'll explore the implications of this further in Bonus Session 3b, but for now we're interested in how that came to be. There are several ways to see this:

1. We were specially created as we are – evolution had no part to play.
2. God took evolved *Homo sapiens* and breathed immortal souls into them.
3. There was a growing spiritual awareness in *Homo sapiens*.

For all three views, there is a point of coming into relationship with God. The image of God is something God gives to each one of us and has nothing to do with our own abilities.

It would be impossible to know the details for any of these possibilities but, whatever happened, at least as far as positions 2 and 3 are concerned, at some point a pair or group of creatures came into personal relationship with God.

Q2: Which of the three "image of God" views, above, fit to which "Who were Adam and Eve?" views on the Briefing Sheet? What do you think of these views, and how they fit the biblical account?

A Random Process?

Dr. Ard Louis said that the word "random" can mean two things:

- In everyday life we use it to mean "purposeless."
- From a scientific way of thinking it means that the tiny details are unpredictable – while the overall process can still be very predictable.

And it's clear that evolution is the scientific type of random process: it acts as a "random optimizer" to find the best solution to a problem.

Read Proverbs 16:33 and Genesis 50:20.

Q1: Keeping in mind the second definition of the word "random," can you think of something in your life that seemed chaotic or random to you at the time but made sense afterwards?

Q2: What do you think of the idea that God might work in creation in a similar way?

Q3: Is it possible that God could be in control of a process like evolution?

Suffering

> “The consequences of the evolutionary process are, admittedly, at times things that cause suffering for individuals even today. A child with cancer may well be seen as one of those side-effects of the fact that DNA copying is not perfect, **it's important that DNA copying not be perfect or evolution wouldn't be possible, but if it results in a cancer arising in a child, isn't that a terrible price to pay**? These are difficult questions to be sure.
>
> Dr. Francis Collins

> “**Why would a loving God allow a tsunami that would kill hundreds of thousands of people?** There are various explanations; I'm not sure that any of them are completely satisfactory. This is one of the toughest questions that believers have to face.
>
> Dr. Francis Collins

Read Genesis 3:1–24.

Q1: How would you define “physical evil”? How is it different from any other kind of evil?

Q2: What are the possible explanations for the existence of “physical evil”?

SESSION 2

How could pain and suffering be part of God's original plan for us in a world he declared to be “good”? There are several possible ways of looking at this:

1. Would physical suffering have been experienced as evil before the fall?
2. Would people have been protected from physical suffering in the Garden of Eden?
3. What if the good world was not meant to be a paradise, but the place where people are made ready for eternal life in the new creation?

Q3: What do you think? Do you think that Adam and Eve, even if they didn't die, would have experienced pain and suffering?

Caring for Creation

Read Psalm 104.

Q1: What does this psalm tell us about the relationship between God, people and creation?

Q2: What have you heard recently about environmental issues?

Read Genesis 1:26–30 and Genesis 2:15.

Q3: What is the status of people in relation to the rest of creation? What did God command them to do?

Q4: What are some ways you can start to care for the environment where you live?

Read Colossians 1:15–20 and Romans 8:19–22.

Q5: How can Christians have hope now and hope for the future?

See page 44 for the "Taking it Further" list.

In-depth Session 2.1: Interpreting Genesis

Preparation for Participants *(optional):* Views on creation

If you had to explain your own views on creation, what would you say?

or

Ask two or three friends, family members or colleagues what they think about how the universe, with our planet and all the life we see on it, came to be.

Glossary	
DNA	The chemical molecule inside every cell of every living thing that carries the instructions for its growth and development.
Intelligent Design	The idea that some parts of living things are too complex to have evolved, coupled with the idea that the information contained in DNA cannot have arisen by a process describable in purely material terms, so providing evidence for "design."
Theistic Evolution (or Evolutionary Creationism)	The belief that God created life through the process of evolution.
Young Earth Creationism	The belief that Genesis should be interpreted as a literal, historical and scientific account, and therefore that God created the world between 6,000 and 10,000 years ago in six twenty-four hour days.

Discussion Topics

Genesis 1

Read Genesis 1.

Q1: What are the most important points in this passage? What does it say about God? About God's relationship with the universe? About God's relationship with people?

Q2: What part does the timescale have to play in thinking about the main points in Genesis 1?

Interpreting Genesis

Read the Briefing Sheet: Views on Genesis 1.

There are many different interpretations of Genesis 1, but the Briefing Sheet outlines three of the most widely-held views. Many of the other views are variations of these.

Q: Which of these views have you come across before?

A Deeper Meaning

View 3 on the Briefing Sheet: Views on Genesis 1 says that Genesis 1 is a piece of literature that describes a real event in non-scientific language.

Q1: Can you give any examples of stories in the Bible that have a deeper meaning?

Q2: Can you give an example of a story being used in the Bible to explain a real event?

Q3: What do you think of the idea that Genesis 1 might also have a deeper meaning?

A Question of Days

How long are the days in Genesis 1:

- Actual twenty-four hour days? (view 1)
- Long periods of time? (view 2)
- Symbolic of God's act of creation? (view 3)

Q1: How many different uses of the word "day" are there in English?

Read Genesis 2:4; Exodus 16:30; Deuteronomy 28:33; Joel 1:15; and 2 Peter 3:8.

Q2: What are the different uses of the word "day" in these passages? Can you think of any more?

Note: The Hebrew word *yom* has the same meanings as "day" in English, but in Hebrew there isn't another word for an extended but fixed period of time (e.g., in English we have "epoch," "era" and "age" as well as "day").

Q3: Look at Genesis 1 again. What do you think the "six days" mean in this passage?

What's the Bottom Line?

Q: Think back to your answers to Q1 in the "Genesis 1" Discussion Topic above. Did any of your discussion in the other sections contradict the core points in Genesis?

See page 44 for the "Taking it Further" list.

In-depth Session 2.2:
An Accident in the Making?

Preparation for Participants *(optional):*
Thinking about evolution

For an introduction to evolution from a Christian perspective, read:

Denis R. Alexander, "Is Evolution Atheistic?"
http://cis.thevirtualchurch.co.uk/assets/files/Resources/Articles/Article-Archive/evolution_atheistic.htm (introductory)

or

Simon Conway Morris, "Extraterrestrials: Aliens like Us?"
http://adsabs.harvard.edu/abs/2005A&G....46d..24M (a paper about evolution)

Glossary	
chromosome	Each DNA strand in a living cell is wound up tightly into a chromosome. Depending on how human chromosomes are visualized in the lab, they can sometimes look like "x" shapes or pairs of striped socks.
genetics	The study of inherited characteristics and the variation of inherited characteristics among populations.
Human Genome Project	The international project to "read" the whole of the human DNA code (the genome).
nihilism	Lack of belief in the existence of morality or meaning in life.

What is Evolution?

1. There is a huge amount of variation in nature – you only need to think of dogs or cats to realize this.

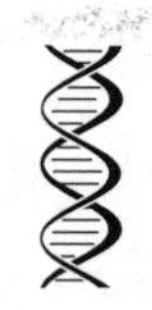

2. Differences in genes (the DNA instruction manual that is in every cell of every living thing) cause a lot of this variation.

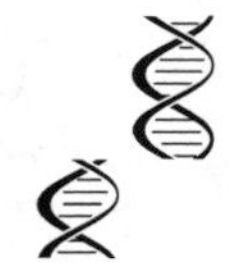

3. The differences are caused by very rare changes in the genes as they are passed on.

Most will not have any effect.

Some will make them sick.

A very few will make them healthier than ever before.

4. Some variations will be more successful than others – certain breeds of cat may have more kittens than others, especially in the wild.

5. These genetically more successful cats will pass on more copies of their genes.

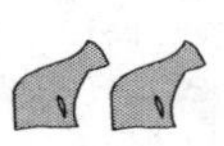

6. Eventually these more successful families will build up a range of genetic differences and become so different that they will form a new species that cannot breed with any animals outside of their own group.

Discussion Topics

 It's a Process

The history that has been painted for us in the documentary is a very long process – the formation of the universe, our planet, and the life on it.

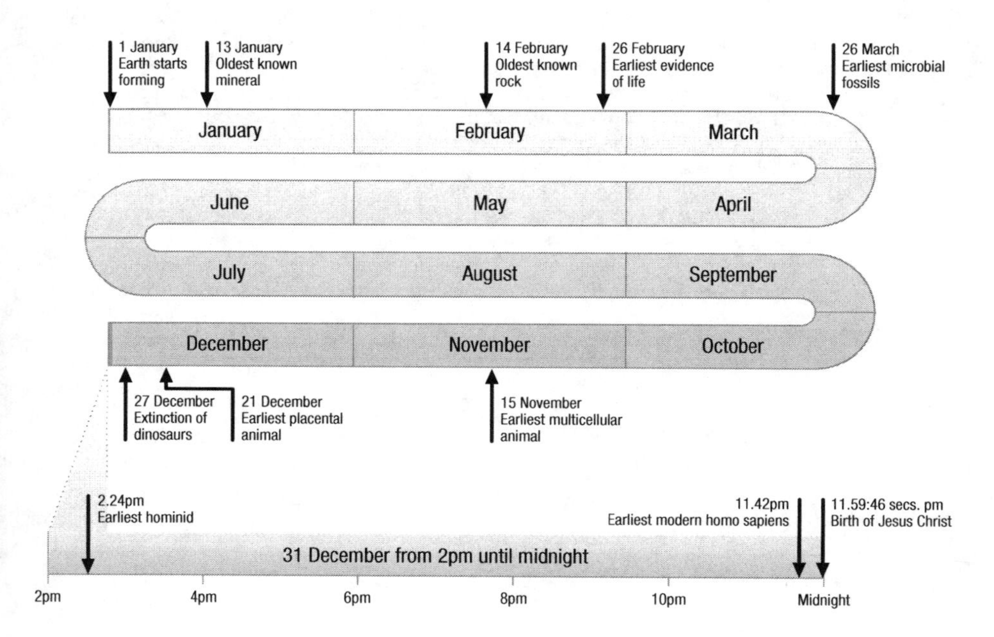

Copied with permission from D. Alexander and R. S. White, *Beyond Belief: Science, Faith and Ethical Challenges* (Lion, 2004), p. 98.

Why would God bother using such a long process when he could clearly create everything in an instant if he wanted to? Some people see a parallel between this process and the long history of the formation of Israel in the Old Testament and the spreading of the news about Jesus in the New Testament.

Q: Do you think this is justified? Think about what the Bible says about God's character and about your own experience.

What a Waste?

" I think one of the biggest questions that many Christians find themselves wrestling with is, **'How can the process of evolution, with all its waste – or at least all its apparent waste – be reconciled with the idea of a loving God, a God who has purpose?'**

– Professor Alister McGrath

Q1: What "waste" do you see in the long process of creation that scientists propose?

SESSION 2

Q2: When you look at creation as a whole, rather than from the point of view of one individual bird, seed, species or star, what are some ways to view the long process besides as "wasteful"? In other words, how could God (and how could we) value species that were extinct long before we came on the scene?

Some more examples:

- The universe needed to be the size that it is to sustain our planet for a long enough period of time for life to develop and remain stable for even just a few thousand years.
- When a star explodes and dies, new elements are made that form new stars and planets. Our bodies were formed from the dust of planet earth, so we are made of stardust.
- Tectonic plates (the earth's outer crust) move and cause earthquakes and volcanoes. These bring up rich nutrients from the earth's core that are vital for sustaining life on our planet.

These scientists see this process as part of God's generosity, or extravagance. The God who "owns the cattle on a thousand hills" (Psalm 50:10) has made us a huge universe and a rich world!

Q3: Do you think it's justified to say that there would be no waste in a long process of creation?

? Animal Death

If God created using a process of evolution, this would mean that the death of many animals was involved. Could this really be part of God's original plan?

Read Genesis 4:3–7; Genesis 9:3; Job 38: 39–41; and Psalm 104:21.

Q1: What is God's attitude to animal death in the Bible? Is it different from human death?

Read Genesis 1:30 and look at the Briefing Sheet: Views on Genesis 1.

Q2: How would different people interpret this verse? Who was eating what (or whom!)?

Q3: Which of these views do you think makes the most sense in the light of God's attitude to animals in the rest of the Bible (Q1)?

A Random Process?

Dr. Ard Louis said that the word "random" can mean two things:

- In everyday life we use it to mean "purposeless."
- From a scientific way of thinking it means that the tiny details are unpredictable – while the overall process can still be very predictable.

And it's clear that evolution is the scientific type of random process: it acts as a "random optimizer" to find the best solution to a problem.

Read Proverbs 16:33 and Genesis 50:20.

Q1: Keeping the second definition of the word "random" in mind, try to think of something in your past life that seemed chaotic or random to you at the time, but made sense afterwards.

Q2: What do you think of the idea that God might work in creation in a similar way?

Q3: Is it possible that God could be in control of a process like evolution?

Is there Purpose in Evolution?

Read the Briefing Sheet: Is there Purpose in Evolution?

" But is this another scientist claiming science can prove God?

" Emphatically I would not say, here's convergence, this is a proof of … anything in fact. **It merely says that the world is structured, and then you must stand back from that position and say, 'Well, why is the world structured in the way it is?'** And if there are claims made by particular religious traditions, are they themselves in any way congruent with those world pictures?

Professor Simon Conway Morris

Q: Do you think this view of the world fits with a Christian belief in a God who intentionally created a world and the people in it? Why or why not?

See page 44 for the "Taking it Further" list.

In-depth Session 2.3:
Evolution, Suffering and the Fall

Preparation for Participants *(optional)*:
Thinking about evolution

As background reading for this session, read:

Graeme Finlay, "*Homo divinus:* The Ape that Bears God's Image"
www.scienceandchristianbelief.org/articles/finlay.pdf

or

Explore the following web page, looking in particular for information about different views on the creation of humans:

The American Scientific Affiliation's "Creation and Evolution" page (different views from a Christian perspective) **www.asa3.org/ASA/topics/Evolution/index.html**

Glossary	
the fall	The account of how people became disobedient to God.
germ cells	Eggs and sperm.
malignant	Bad or harmful, often used with regards to cancer; the opposite of benign (harmless).
mutation	A change in the DNA code that happens during the life cycle of a living thing. Mutations can be caused by a toxic chemical or other environmental disturbance, or by a mistake in copying the DNA when new cells are made.
somatic cells	All the cells in the body except eggs and sperm.

Discussion Topics

Human Evolution

Dr. Francis Collins outlined the evidence for human evolution in Chapter 2 of Part 2 of the DVD.

If he is right, this immediately raises two questions: "Who were Adam and Eve?" and "How are we special compared to other animals?"

Read Genesis 2:4–25.

Q1: What are the most important points in this passage? What does it say about God? About God's relationship with the universe? About God's relationship with people?

Read the Briefing Sheet: Views on Genesis 2 and 3, Who were Adam and Eve?

Q2: Which of these views do you think could fit best with the biblical and scientific account?

So how are we special compared to other animals? The Bible says that we are made "in God's image." Bonus Session 3b will explore the implications of this further, but for now we are interested in how that came to be. There are several ways to see this:

1. We were specially created as we are – evolution had no part to play.
2. God took evolved *Homo sapiens* and chose them to bear his image by divine fiat.
3. There was a growing spiritual awareness in *Homo sapiens*.

For all three views, there is a point of coming into relationship with God. The image of God is something God gives to us and has nothing to do with our own abilities.

It would be impossible to know the details for any of these possibilities but whatever happened, at least as far as positions 2 and 3 are concerned, at some point a pair or group of creatures came into personal relationship with God.

Q3: Which of the three "image of God" views, above, fit to which "Who were Adam and Eve?" views on the Briefing Sheet? What do you think of these views and how they fit the biblical account?

Q4: There are many views on the interpretation of Genesis 2 (discussed above). Did any of your discussion complement or contradict the main points of Genesis 2 that were discussed in Q1?

The Fall

" Our fall, therefore, is … [our] falling out of God's purposes for humanity, a refusal to be what God made us to be, a turning away of a summons and a calling to reflect God back to himself in our dealings with him and with one another.

Tom Smail, *Like Father, Like Son: The Trinity Imaged in our Humanity* (Paternoster, 2005), p. 213

Read Genesis 3:1–24.

Q1: How were Adam and Eve disobedient? What are the consequences of their sin? What can we learn about God from his response?

The Bible talks about three types of death:

- Physical death (Genesis 25:8)
- Spiritual death (Colossians 2:13)
- Eternal spiritual death, or second death (Matthew 10:28)

Q2: Which type of death do you think resulted from Adam and Eve's disobedience? When or how did it happen?

Read the second half of the Briefing Sheet: Views on Genesis 2 and 3, Views on the fall.

Q3: Which of these views do you think could fit best with the biblical account?

Suffering

" The consequences of the evolutionary process are, admittedly, at times things that cause suffering for individuals even today. A child with cancer may well be seen as one of those side-effects of the fact that DNA copying is not perfect, **it's important that DNA copying not be perfect or evolution wouldn't be possible, but if it results in a cancer arising in a child, isn't that a terrible price to pay**? These are difficult questions to be sure.

Dr. Francis Collins

" **Why would a loving God allow a tsunami that would kill hundreds of thousands of people?** There are various explanations; I'm not sure that any of them are completely satisfactory. This is one of the toughest questions that believers have to face.

Dr. Francis Collins

Q1: How would you define "physical evil"? How is it different from any other kind of evil?

Q2: What are the possible explanations for the existence of "physical evil"?

Q3: In Genesis 1:31 God says that the world is "very good." Do you think this could include the death and extinction of many species?

SESSION 2

How could pain and suffering be part of God's original plan for us in a world God declared to be "good"? There are several possible ways of looking at this:

1. Would physical suffering have been experienced as evil before the fall?
2. Would people have been protected from physical suffering in the Garden of Eden?
3. What if the good world was not meant to be a paradise, but the place where people are made ready for eternal life in the new creation?

Q4: What do you think? Do you think that Adam and Eve, even if they didn't die, would have experienced pain and suffering?

Read Revelation 21:1–4.

Q5: What is God's ultimate purpose for creation?

Taking it Further

Websites and articles to download on creation:

Bob White, "The Age of the Earth"
www.st-edmunds.cam.ac.uk/faraday/resources/Faraday%20 Papers/Faraday%20Paper%208%20White_EN.pdf

Ernest Lucas, "Interpreting Genesis in the 21st Century"
www.st-edmunds.cam.ac.uk/faraday/resources/Faraday%20 Papers/Faraday%20Paper%2011%20Lucas_EN.pdf

Graeme Finlay, "*Homo divinus:* The Ape that Bears God's Image"
www.scienceandchristianbelief.org/articles/finlay.pdf

The American Scientific Affiliation's "Creation and Evolution" page (different views from a Christian perspective) **www.asa3.org/ASA/topics/Evolution/index.html**

Answers in Genesis (Young Earth Creationism): **www.answersingenesis.org**

The Discovery Institute (Intelligent Design): **www.discovery.org**

Books on creation:

Deborah B. Haarsma and Loren D. Haarsma, *Origins: A Reformed Look at Creation, Design and Evolution* (Faith Alive Christian Resources, 2007). This is a very approachable introduction to the area of science and faith, and creation. It examines the origins of the universe and living things. Further material is available on **www.faithaliveresources/origins**, including questions that can be used by a small reading group. The book covers a range of views.

Norman Geisler (ed.), *The Genesis Debate: Three Views on the Days of Creation* (Crux Press, 2001). Three pairs of authors present different views on the days of creation: the 24-hour view, the day-age view, and the framework view, and respond to each other's writing.

Paul Nelson, Robert C. Newman and Howard J. Van Till, *Three Views on Creation and Evolution* (ed. John Mark Reynolds and J. P. Moreland; Zondervan, 1999). Proponents of Young Earth Creationism, Old Earth Creationism and Theistic Evolution each present their different views, explain why the controversy is important and describe the interplay between their understandings of science and theology. Various scholars critique each view.

David Wilkinson, *The Message of Creation* (The Bible Speaks Today; IVP, 2002). This is a very thorough look at the themes of creation throughout the Bible, beginning with Genesis 1–3. The following sections deal with: the songs of creation that praise our Creator God; Jesus' relationship to creation; the lessons the writers of the Bible teach using creation; and the new creation.

Ernest Lucas, *Can We Believe Genesis Today?* (IVP, 2005). This book explains how scholars have interpreted Genesis 1–11, historically and in the light of modern science. Lucas looks at various interpretations, noting the problems with each and giving sources for further reading. He highlights his own view, that mainstream science and the Bible are compatible.

Denis Alexander, *Creation or Evolution: Do We Have to Choose?* (Monarch, 2008). This is an in-depth look at all the questions concerning creation and evolution, from the perspective of Theistic Evolution.

Darrel Falk, *Coming to Peace with Science* (IVP, 2004). Darrel Falk sympathetically picks his way through the various theological arguments on all sides of the debate and comes to the conclusion that Christianity is compatible with evolutionary biology.

Bonus Session 2b: The Environment

Preparation for Participants *(optional)*: The environment

Read one of the articles suggested on the "Taking it Further" list, or the two Briefing Sheets for this session.

or

Keep an eye (or ear) out for stories in the media about the environment. What are the issues? Who do they affect? What solutions are suggested?

Discussion Topics

Environmental Issues

Q1: What have you heard, read or watched about the environment or environmental issues recently?

Many of the issues we face are interconnected:

Water – provision of clean water for a growing population and problems caused by drought or floods

Climate change – a small increase in the overall temperature of the earth's atmosphere has a huge effect on the weather

Population[i] – The population of the world has almost doubled in the last 40 years. How will everyone have enough food and water? (World population in 2005: 6,514,751; and in 1965: 3,342,771 – an increase of x1.95.)

Soil degradation – Overgrazing and deforestation mean that soil is washed from exposed land

Destruction of the places where living things usually grow (habitat loss), and a reduction in the number and variety of living things growing in the world **(reduced biodiversity)**

Q2: Can you map out how these things might be related?

Q3: How seriously do you think we should take these issues, especially climate change?

Environmental Issues *(optional)*

Watch the bonus interview: 2.1 John Houghton

The IPCC was an international panel set up by the by the World Meteorological Organization (WMO) and the United Nations Environment Programme (UNEP) to assess the scientific evidence for human-induced climate change.

Sir John Houghton was the chair or co-chair of the scientific assessment working group of the IPCC from 1988 until 2002.

Q4: What does the story in this clip say about John Houghton's experience as a scientist and a Christian?

Read the Briefing Sheet: Climate Change Questions.

Some do not agree that people cause climate change. The Briefing Sheet outlines some of these views, as well as a response.

God's Purpose for Creation

Read Psalm 98:4–9; Psalm 19:1–4a; and Romans 1:20.

Q1: What do these passages say about the non-human creation? How does it glorify God?

Q2: Are any of the things discussed in Q1 done for our benefit?

Read Genesis 1:26–30 and Genesis 2:15.

Q3: What is our status in relation to the rest of creation? What does God command us to do?

The Putrefied World

Read the passages below and identify the chain of broken relationships in creation:

Genesis 3:8–11

Genesis 3:12–13

Genesis 3:15–19

Genesis 4:3–9

Genesis 6:11–12

Watch the bonus interview: 2.2 John Houghton

The Purified World

Read Colossians 1:15–20.

Q1: What was and is Jesus' role in creation?

Watch the bonus interview: 2.3 Alistair McGrath

Read Romans 8:19–23.

Q2: What will ultimately happen to the non-human creation?

Read 2 Peter 3:10–13.

Q3: If creation is renewed, what might the fire in this passage refer to?

What Do We Do?

Q: With all the environmental problems that we face how can we have hope now, and for the future?

Watch bonus interviews: 2.4 Ian Hutchinson, 2.5 Catherine Cutler, 2.6 John Houghton

Activities

1. Cook a sustainable, local and seasonal fellowship meal

If food is grown locally it usually will not have been transported so far, especially if bought from a local store. (Some of the larger chains will buy produce locally then ship it far away to have it packaged for less than it would cost locally!) From a purely culinary point of view, food bought from a local producer usually tastes much better. It can also be cheaper.

2. Make your garden bee and butterfly friendly

Make sure there is a source of water (a bird bath with a few twigs will do). Plant pollen- and nectar-rich plants. Many modern varieties of flowers don't produce very much pollen, but old varieties and wild roses work well. In order for plants to produce good nectar they need good soil, so making sure that it is just slightly acidic and has the right nutrition is important. A good way to do this is to dig in lots of fertilizer or mulch. Try not to use herbicides and pesticides, but if you must do check to make sure they are bee friendly. Plant flowers together instead of in small islands, and try to have plants that flower at different times of the year. Learn about local species habitats. For butterflies, plant the plants which they like to lay eggs on (and that become food for caterpillars). Plant shrubs for shelter and, if possible, some fruit-bearing trees or bushes.

3. Discussion

Have you ever heard a person being referred to as a "consumer"? What is the difference between the words "person" and "consumer"? What does this say about our contemporary outlook? What's God's vision?

- You could make a list of things people are called in contemporary culture (e.g., person, consumer, agent, character, actor, steward).
- What do these things mean? How do they affect the way we view ourselves? What is God's view?

4. Go for a walk

The aim is to get to know what's living and growing in your area. It would be useful to look up what you'd be likely to find in advance, so that you can identify what you see. It is also helpful to have interesting facts on hand about the behavior and growth of the things you might see – once you have one interesting hook, it makes learning much easier.

5. Bird houses

Buy or make bird houses. If you don't have a yard, see if your local school, church, nature reserve or park would appreciate them.

6. Plant a tree

Trees not only take carbon out of the atmosphere and store it for many decades, they also support biodiversity. Trees in some areas can have as many as a thousand different species living on them! It would be useful to do some research about local tree species to find out what will do well in the area where you want to plant. As with the bird houses, if you don't have a yard see if your local school, church, nature reserve or park would appreciate some trees being planted.

7. Learn about local natural history and ecosystems

Look for activities in your area where you can do this.

8. Campaign

If you don't have the facilities to recycle, take public transportation or cycle in your area, **write to local politicians** asking to increase (or introduce) funding for these things in the local area.

Taking it Further

Websites:

A carbon footprint calculator and carbon offset scheme
www.climatestewards.net

Resources from a Christian conservation group
http://en.arocha.org/home

Resources and articles
www.jri.org.uk

Articles and courses
www.ausable.org

The UK "Ecocongregation" scheme
www.ecocongregation.org

Environment resources from the American Scientific Affiliation
www.asa3.org/aSa/topics/environment/index.html

Evangelical Environment Network
www.creationcare.org

Articles to download:

John Houghton, "Why Care for the Environment"
www.st-edmunds.cam.ac.uk/faraday/resources/Faraday%20 Papers/Faraday%20Paper%205%20Houghton_EN.pdf

John Houghton, "Global Warming, Climate Change and Sustainability"
www.jri.org.uk/brief/Briefing_14_3rd_edition.pdf

"Climate Change Controversies: A Simple Guide" from The Royal Society, 2007
http://royalsociety.org/Climate-change-controversies

Robert White, "A Burning Issue: Christian Care for the Environment"
www.jubilee-centre.org/document.php?id=53

Books:

Hilary Marlow, *The Earth Is the Lord's: A Biblical Response to Environmental Issues* (Grove Books, 2008). Looking at a range of texts and themes in the Old and New Testaments, this study shows how the whole of the non-human created order is included in the biblical vision of God's restoration. It includes questions for reflection and points to resources for practical action.

Colin A. Russell, *Saving Planet Earth: A Christian Response* (Authentic, 2008). This is a very introductory level book, rich in facts about the environment, theological reflection on the Bible and practical suggestions. It addresses some questions which might be more specific to the evangelical wing of the church.

Dave Bookless, *Planetwise: Dare to Care for God's World* (IVP, 2008). This is an introductory level book setting out what the Bible says about why and how Christians should care for God's earth. It is full of practical illustrations and suggestions as well as biblical material. It also includes follow-up questions and resources.

Martin J. and Margot R. Hodson, *Cherishing the Earth: How to Care for God's Creation* (Monarch, 2008). This is a stimulating and inspiring Christian response to environmental issues from a scientist and a theologian. It includes chapters on the science of global warming and on the effect on the world's poor and challenges our attitudes and lifestyles.

Nick Spencer, Robert White and Virginia Vroblesky, *Christianity, Climate Change, and Sustainable Living* (Hendrickson, 2009). Spencer and White look at the science behind climate change and the biblical imperative behind Christian engagement with environmental issues. They diagnose modern cultural problems leading to climate change and include practical suggestions for ways to integrate care for creation at different levels of life.

“What on earth am I doing?” A personal lifestyle audit

(Adapted with permission from Ruth Valerio, author of *L is for Lifestyle: Christian Living that Doesn't Cost the Earth* [IVP, 2008].)

Being a Christian should challenge us to face the issues of our world around us and do something about them. These questions provide a measure for checking out your lifestyle and thinking with reference to the environment. Your answers and scores should be a stimulus for discussion and action – mark yourself as honestly as you can! Use the definitions below to clarify the questions. Please tick a box for each question.

I Buy:	I do it	I think about it	It doesn't cross my mind
Environmentally friendly laundry detergent			
Items with less packaging (whenever possible)			
Items with less transport miles (when aware)			
Recycled paper / envelopes / toilet paper / paper towels			
I recycle:			
Newspapers / waste paper			
Glass			
Aluminum or steel cans and plastic			
Garden waste			
Kitchen waste			
Clothes / books			
I make a point of using:			
Local stores instead of out-of-town supermarkets			

Local farmer's markets and farm shops			
Public transportation / carpooling			
A bike instead of a car			
A car with a small engine			
Energy-saving light bulbs			
Lights / electrical equipment and turn off when finished (not on stand-by)			
Produce which I have grown myself			
Gas and electricity from a green energy provider (where available)			
I support:			
Local conservation groups			
National environmental organizations			
Birds, by providing food in my garden and putting bells on any cats I own			
Local wildlife by gardening organically			
Add up your scores in their columns. Each point is worth:	**2**	**1**	**0**
Grand total	+	=	

SESSION 2

Scores and Definitions

0–16	Being a Christian doesn't impact your lifestyle or thinking about these issues much. Choose an issue which interests you and discover how you can make a difference.
17–32	You're thinking about making a difference, but getting around to it remains a challenge. It's time to do those things you've been putting off!
32–48	Your lifestyle reflects that you've made changes. Challenge yourself to find out more and keep going!

Environmentally friendly means being sensitive to the need to reduce the use of natural resources, considering pollution and the amount of energy used by producing or using a product.

Transport miles refer to the mileage covered by an item from the producer of the raw ingredients to the shop floor. For example, a locally grown potato may travel to a washing center and then to a distribution center before it reaches your local superstore, however the local market will sell it dirty direct from the farm! More transport is used, and therefore more congestion and pollution are produced, by shopping at superstores.

Recycling is the idea of using materials again. If an item cannot be re-used in its present form, it can be broken down and the materials used again. This process uses far less energy and fewer natural resources than using raw materials each time.

Carpooling makes use of spare seats in cars when two or more people are travelling to the same destination at the same time.

Farmers markets are markets where local producers can sell their goods direct to the customer. In the UK, for example, they must come from within a 30-mile radius and the stall has to be staffed by the actual producer. The produce is not only fresher but often also contains few chemicals. Less packaging and transportation are required, which means there is less waste and fewer road journeys. Farmers markets also encourage people to try home-produced, regional specialities.

Bicycling is a far more environmentally friendly means of transportation than driving. For example, a bicycle can be pedalled for up to 1037 km on the energy equivalent of a quarter of a gallon of gas (nearly 300 mpg). In addition, a regular adult cyclist on average exhibits the fitness levels of someone ten years younger.

Session 3: Is There Anybody There?
Freedom to Choose

Preparation for Participants *(optional):*
Who am I?

Ask two or three friends, family members or colleagues to think of two things about themselves (physical characteristic, talent, interest, like/dislike, etc.):

1. Something that they believe they inherited from a parent. (Although unless it's an obvious physical characteristic it's difficult to figure out whether it was learned from them or actually inherited – you'll just have to guess!)
2. Something that they think came from another source. (E.g., something learned from people around them, something in their physical or cultural environment, a spiritual experience, a new opportunity, or perhaps something they developed by themselves.)

or

Ask a relative or family friend what things they think that you inherited from parents and what things are unique to you, and perhaps sometimes surprising?

Glossary	
altruism	Unselfish concern for others (which may involve self-sacrificial acts).
cell	The units that make up a living thing. Animal cells consist of a membrane enclosing whatever parts that particular type of cell needs to do its job. A fat cell contains fat, a bone cell contains a hard substance, a red blood cell contains a substance that carries oxygen around the body, and the long spindly nerve cells are able to pass electrical signals along their length.
emergence	The idea that complex structures have properties that you couldn't predict if you looked at their individual parts.
neurons	The "nerve cells" that carry messages in the nervous system and the brain.
neuroscience	The study of the brain and nervous system.
reductionist	Someone who thinks that you can explain everything by reducing it to its most basic physical properties.

Discussion Topics

The "God Spot"

Dr. Alasdair Coles says that some neuroscientists (*not* including himself) believe that spiritual experiences are just a side-effect of the brain's normal function.

Q: The scientists interviewed and many others (of many faiths and none) have rejected the idea of a "God spot" in our brains. Do you think they're right? Use the two quotes below to help you.

" [There's] a place in the brain for everything. You know there's a Jennifer Aniston spot and there's a hamburger spot in my brain and in yours. Anything you know anything about, **anything you have any bunch of beliefs about, there's got to be something in your brain that's holding those.**

Professor Daniel Dennett, philosopher[ii]

" If someone wanted to come along and link me up to electrodes while I was praying or while I was in worship and found that my brain patterns were slightly different, then that wouldn't be a great surprise to me. I think spiritual experience is real and therefore there should be a way of looking at that in terms of the physicality of the brain. **But just to look at those brain patterns and to say that that is all that spiritual experience is seems to me to be mistaken.**

Dr. David Wilkinson

Whole Persons

Dr. Alasdair Coles says that we are not simply machines, and neither are we disembodied souls with no real connection to our physical bodies. A person's body, mind and soul are interdependent.

With this in mind, read Mark 12:28–30.

Q1: How could you obey this command?

Read the account of Jesus' resurrection appearance in Luke 24: 36–44 and his promise that we will be resurrected in John 5:28–29.

Our physical bodies are clearly important to God – not just the part of us that may be purely spiritual. There is somehow continuity between our physical bodies here and now and our resurrected bodies.

Q2: What ideas have you heard or read about the body and spirit or soul? How do those ideas compare to the account in the passages you have just read?

Choices, choices ...

Professor Bill Newsome says that although there are obviously some things that we can't do (not all of us can play golf like Tiger Woods, for example), we do have the ability to make meaningful choices.

Read Joshua 24:14–15, keeping in mind this idea of making choices.

Q3: What does this passage say about our ability to choose?

Q4: Think of a conscious choice that you made that changed the course of your life.

An Ethical Toolkit

To tackle ethical issues from a Christian perspective you first need to construct a moral framework, or "ethical toolkit," from biblical principles.

Q: What moral principles can you draw from the following Bible passages?

Genesis 1:26–27

Galatians 3:26–29; Romans 12:4–8

Mark 12:31; Philippians 2:3–4

Deuteronomy 10:18

Matthew 25:31–46

Psalms 127:3–5

Using Your Ethical Toolkit: Cloning[iii]

Cloning (often called "reproductive cloning") is making a genetic copy of a person – the process would be like taking a cell from someone's body and using it as a seed to grow a new person.

Q1: Some babies born naturally are genetic clones of each other – identical twins! Thinking about any identical twins that you have met, what can you tell about how much our genes determine our different characteristics?

Q2: Why do you think someone might want to artificially "clone" a person?

Read the case study on page 63-64 and answer the following questions, bearing in mind your "ethical toolkit."

Q3: Can you anticipate some of the consequences of cloning this child? How might it affect the resulting clone, their family, and the rest of society?

Q4: Can you anticipate some of the consequences of people cloning themselves and bringing these clones up as their own children? On the children? The parents? Society?

The procedure of cloning poses huge medical risks:

- A very high miscarriage rate.
- The cell from the cloned person is "old," and the new child made from that cell will inherit that age (including cancer risks), rather than having their "clock" re-set by the process of egg formation. This is a process that biologists don't understand fully, so it is difficult to predict the results or reduce the risk.
- In the process of making eggs and sperm, certain genes are switched on and off. A clone whose "parent" is a fully developed cell may not inherit the correct switching required for normal development.

Q5: Think back to the motivations for cloning a child discussed in Q2. Do you think any of these would justify taking the risk of cloning a child?

Taking it Further

Websites and articles to download:

John Bryant, "Don't My Genes Determine My Behaviour?"
www.eauk.org/resources/idea/bigquestion/archive/2005/bq9.cfm

Denis Alexander, "Cloning Humans – Distorting the Image of God?"
www.jubilee-centre.org/document.php?id=32&topicID=0

Ethics resources from the American Scientific Affiliation
www.asa3.org/ASA/topics/ethics

The International Christian Medical and Dental Association
www.icmda.net

Center for Bioethics and Human Dignity
www.cbhd.org

Books:

John Bryant and John Searle, *Life in Our Hands: A Christian Perspective on Genetics and Cloning* (IVP, 2004). With an eye to the practical application of new technologies, Bryant and Searle lay out the ethical dilemmas facing biological scientists and explore the theological implications. They outline the ethical position that they have reached on each issue, but not before showing the various positions that Christians take and emphasizing how difficult it can be to decide in matters that affect life and death.

Tony Watkins (ed.), *Playing God: Talking about Ethics in Medicine and Technology* (Authentic/Damaris Publications, 2006). *Playing God* tackles ethical issues in a different way, following the Damaris route of using films and books – including a television medical drama, Jodie Picoult's *My Sister's Keeper,* the writings of Isaac Asimov, Margaret Atwood's *Oryx and Crake,* and the ethics of Peter Singer – to discuss the topics. This is an easy but thought-provoking read and could be the basis for a group study.

Pete Moore, Babel's Shadow: *Genetic Technologies in a Fracturing Society* (Lion, 2000). Written by a biologist who is an experienced science writer and ethics lecturer, this book covers the broad issues involved in decisions concerning genetics and its medical applications. Although written before the completion of the Human Genome Project, the principles here still apply and this book is an excellent and very readable introduction to the topic.

John Wyatt, *Matters of Life and Death: Today's Healthcare Dilemmas in the Light of Christian Faith* (IVP, 1998). In writing this book John Wyatt draws on his experience as Professor of Neonatal Paediatrics and Consultant Neonatal Paediatrician at University College London. This is a very thorough introduction to the issues and draws on a lot of biblical material. It covers reproductive technology, fetal screening, genetics, abortion, neonatal care and euthanasia.

Case Study[iv]

The fog on the highway was exceptionally dense as the Robinson family drove towards London on November 20th, 2011. Their only child Susan, aged four, was playing happily with her dolls on the back seat. After years of unsuccessfully trying for a baby the Robinsons had eventually decided to use *in vitro* fertilization to have Susan, so she was especially cherished. Her long eyelashes and dimples were the spitting image of her mom, whereas even at that young age her long limbs held great promise of future athletic prowess, or so her proud father liked to think.

Suddenly a pile-up loomed out of the fog in front of them. Mr. Robinson slammed on the brakes. His quick responses prevented their car from diving into the mangled heap of wrecked cars ahead, but unfortunately the truck driver behind was not so alert, sliding into their rear with a sickening thud. Seconds later the shocked parents found themselves clutching Susan's lifeless form as they huddled on the shoulder, waiting for help to arrive.

Minutes later, after a short but fevered discussion, Mrs. Robinson called CLON777 on her cell phone and as the fog began to clear a helicopter landed in a nearby field, CLONE-AID emblazoned across its fuselage. A white-coated medical technician leapt from the helicopter and was soon taking tiny skin samples from Susan's limp body. Minutes later the samples were being stimulated in a nearby CLONE-AID laboratory to establish cell cultures.

Several months went by while the Robinsons grieved for little Susan, but finally they could contain themselves no longer. They wanted a replacement Susan and they wanted her now. Fortunately Mrs. Robinson already had viable eggs frozen down as a result of her cycle of *in vitro* fertilization. The great day came. In the CLONE-AID laboratory, with its picture of Dolly the sheep proudly displayed on the wall, the process of "nucleus transfer" began. A nucleus was removed from one of Susan's cultured skin cells. This single nucleus contained the cell's DNA with its genetic instructions to build a new Susan. Carefully the nucleus was placed in a small dish with one of Mrs. Robinson's eggs from which the nucleus had been removed. A small electric current was zapped through the cell suspension and the nucleus fused with the egg cell to produce a tiny embryo. This procedure was repeated multiple times to generate several embryos that were carefully screened over the next few days to check for any abnormalities before one of them was implanted in Mrs. Robinson. Nine months later the Robinsons held in their arms a pink and gorgeous looking "replacement Susan," complete with dimples, prominent eyelashes and long limbs.

Cloning explained:[v]

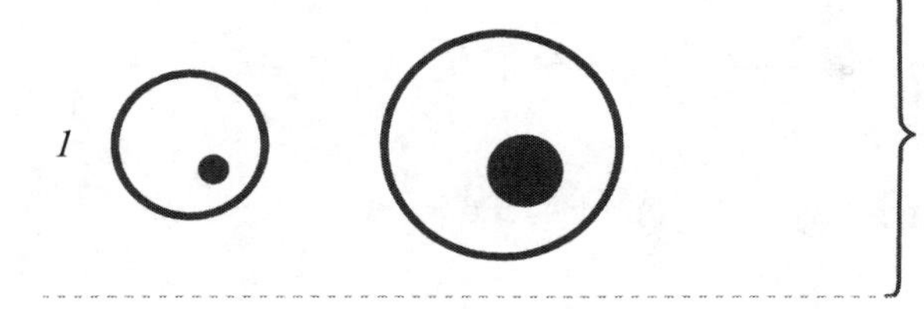

An adult cell (left) and egg cell (right). The DNA of both cells is contained in a structure called the nucleus, represented by the solid black circles. The adult cell contains two sets of DNA; the unfertilized egg contains one.

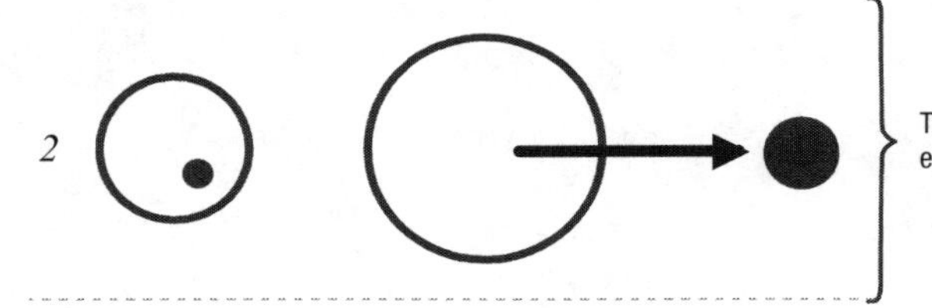

The nucleus is removed from the egg cell.

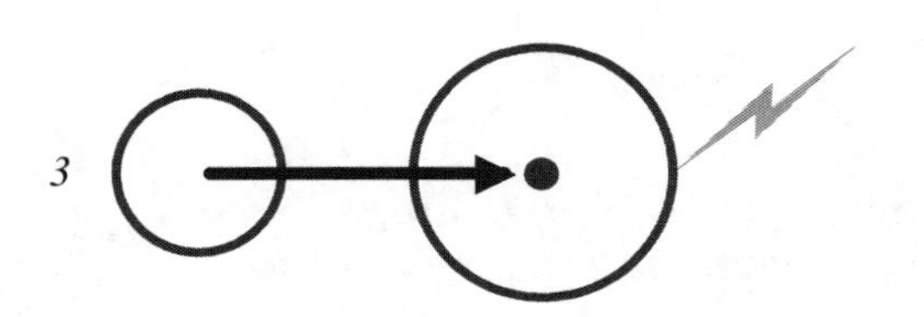

The egg cell nucleus is replaced with a nucleus from the adult cell. The new cell-nucleus combination is stimulated to divide (as a normally fertilized egg would), often by a very brief electric shock. If an embryo is formed it is placed in the womb in order to establish a pregnancy. It will be a genetic copy of the adult from which the nucleus (containing the DNA) was taken.

Bonus Session 3b:
Is There Anybody There?
Thinking about Human Identity

Preparation for Participants *(optional):*
Ethics

Read one of the articles from the "Taking it Further" section for this session.

or (more introductory)

Explore one of the websites recommended in the "Taking it Further" section for Session 3, looking for information on the "Image of God."

Discussion Topics

The Image of God

Read Genesis 1:26–28 and Genesis 2:15–17.

Q1: What do you think the phrase "made in the image of God" means? What is it about us that is different from other living creatures?

Q2: How would you define human uniqueness in a secular context? Is it possible?

Watch the bonus interview: 3.1 David Wilkinson

Glossary	
The Near East	A term archaeologists and historians use for the Middle East.

Q3: Some of the definitions of "the image of God" or human uniqueness that you came up with above might exclude some people. How is the description David Wilkinson gives more inclusive?

Read Genesis 9:5–6 and James 3:7–10.

Q4: What are the consequences of being made in the image of God? How does that affect you?

The Beginning of Life

Watch the bonus interview: 3.2 John Bryant

Read the Briefing Sheet: When Does Human Life Begin?

These are the most commonly held views on the status of embryos and the arguments that are used to support them.

Q: What do you think of the different views laid out here?

Genetic Testing

Watch the bonus interview: 3.3 Denis Alexander

Q1: What might be the arguments for and against testing for genetic diseases, and the possible destruction of affected embryos (this could happen by testing IVF embryos and deciding which ones to use)?

Q2: Who else do you think is affected by genetic testing, in addition to the embryo?

Enhancement

Watch the bonus interviews: 3.4 John Bryant, Francis Collins and John Polkinghorne

Glossary	
cognitive ability	The ability to think and experience things.
transhumanism	The idea of enhancing human abilities with technology in extreme ways.
preimplantation genetic diagnosis	DNA testing of IVF embryos.

Examples of some enhancement technologies:

1. Genetic modification, or GM (e.g., "gene doping" in athletes) *(a speculation for the future):*
   - to cause an effect in an individual
   - or to change the genes in their eggs or sperm so that the future generation is affected.
2. Prosthetics or implants (e.g., an artificial leg, brain implant, living tissue implant) *(many present realities)*
3. Chemicals or drugs (e.g., to enhance memory) *(present realities or will be with us soon)*

Q1: Look at the list of potential treatments below and put them on the spectrum, deciding where they should be on the line from healing to enhancement.

1. Vaccination
2. Caffeine tablets
3. A memory implant (in a normal functioning person)
4. A third arm
5. GM to treat muscular dystrophy[vi]
6. Drugs to enhance concentration
7. GM to make someone taller[vii]
8. Glasses

Healing — Unacceptable enhancement

Q2: What does enhancement focus on? How might that affect the enhanced individual and his or her relationship with the rest of society?

SESSION 3

Read Luke 4:18–19; 17:11–14.

Q3: What different effects does Christian healing have on society?

Q4: Do your answers to Q2 and Q3 change the way you would approach Q1?

Emergence

Watch the bonus interview: 3.5 Bill Newsome

Q1: Can you think of any emergent properties in daily life?

Watch the bonus interview: 3.6 Bill Newsome

Emergence can mean two things:

1. That the whole has a different, or higher, property than its parts. These properties can be defined scientifically, and this sort of emergence is not at all controversial (e.g., a table compared to a piece of wood – this is "epistemological emergence").
2. That "emergent phenomena" have real significance in the world. For most people this is not an important question for a table, but an important question for us is: "Does human emotion have more meaning than the firing of neurons?" This type of emergence is impossible to prove scientifically. (This is "ontological emergence.")

Q2: What do someone else's emotions mean from these two points of view?

Q3: What are the implications of these two views for our ability to make choices? Which do you think makes the most sense?

 **Watch the bonus interview: 3.7 Bill Newsome**

Taking it Further

See also the list from Session 3.

Articles to download:

Michael Poole, "Reductionism: Help or Hindrance in Science and Religion?"
www.st-edmunds.cam.ac.uk/faraday/resources/Faraday%20 Papers/Faraday%20Paper%206%20Poole_EN.pdf

John Bryant, "Ethical Issues in Genetic Modification"
www.st-edmunds.cam.ac.uk/faraday/resources/Faraday%20 Papers/Faraday%20Paper%207%20Bryant_EN.pdf

Books:

Pete Moore, *Enhancing Me: The Hope and Hype of Enhancement* (Wiley, 2008). While this was not written with the intention of giving specifically Christian teaching, it is a very readable and informative look at the latest enhancement technologies. It clearly separates the possible from the currently impossible and is in full color with lots of pictures.

Briefing Sheet Session 1:
Beyond Reason?

CHAPTER 1

The question we are faced with in society today is: **HAS SCIENCE DONE AWAY WITH FAITH?**

But many Christians who are also scientists have no problem reconciling their science and their faith.

The first scientists in the Western world were actually Christians.

The conflict we hear about in the media was mainly stirred up by Victorian scientists who wanted to get rid of the clergy who were involved in science.

One way of making sense of the difference between science and religion is to think of a boiling kettle.

How is it boiling? Because the element is heating the water.

Why is it boiling? Because I want some tea.

In the same way, science asks **"how?"** and religion asks **"why?"**

CHAPTER 2

These scientists believe that God created the universe. The **"how"** is answered by science – they see evidence that the universe started with a **"Big Bang."**

We can make sense of the world using science and mathematics.
E.g., it was even possible to predict something as strange as antimatter using mathematics – several years before there was any physical proof for it.
But **why** is it like that? Is it because the universe was created by an intelligent being?

GOD OF THE GAPS

There is a lot about the universe that is still unexplained.

Are the unexplained parts proof for God? No, because scientists might find an explanation for these things in the future.

If we say these gaps are evidence for God, then if the evidence disappears, where is God?

It's far better to look at the whole picture and say, "What an amazing creator!"

CHAPTER 3

For example, the chances of the Big Bang resulting in a universe, and our planet that can support life, are very small. Even those scientists who are atheists have been surprised by this.

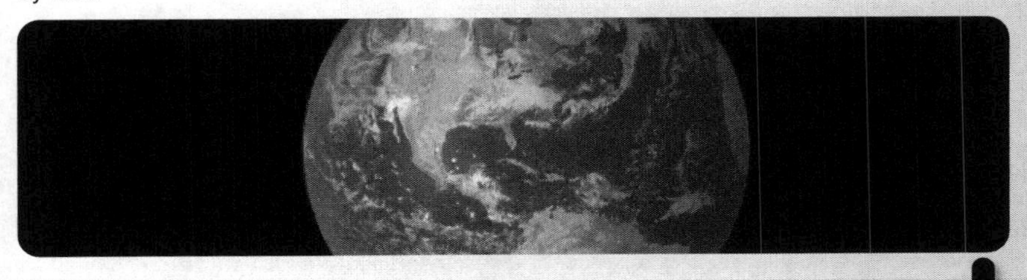

CHAPTER 4

DO WE LIVE IN A MULTIVERSE?

Some scientists explain the existence of life on earth by saying there are many universes, so it's not surprising that one has turned out like ours.

But there is no evidence at all for these other universes. Neither will it mean, if they do exist, that they are outside God's creation.

Science can only answer certain questions – it doesn't tell you anything about the meaning of life.

Faith isn't threatened by science. Why should we worry about finding out more about the world God created? In fact, the more we find out, the more we can be amazed by what God has made.

Briefing Sheet Session 1:
The Big Bang

How did the universe start?

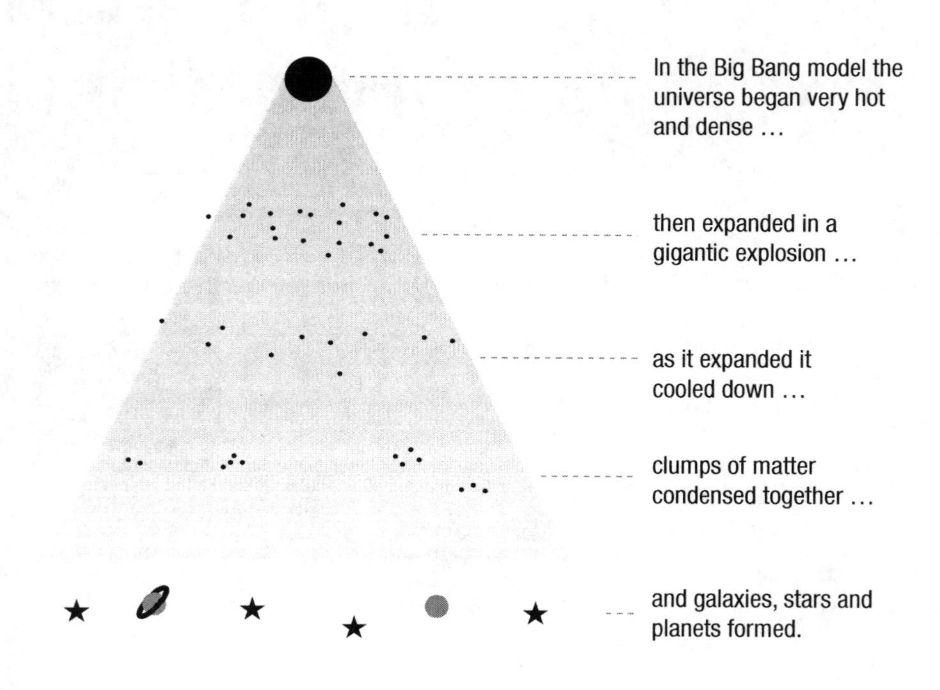

In the 1920s, Edwin Hubble discovered that the universe is still expanding. Astronomers detect that all the galaxies are moving away from us, and that more distant galaxies are moving away faster. This relationship between distance and speed means that the universe as a whole is expanding. In the 1990s, astronomers found that the expansion rate is not constant but is speeding up over time. Using the current best measure of expansion rate and how it changes over time, astronomers calculate that the universe itself must have begun about 13.5–13.9 billion years ago.

In 1965, Arno Penzias and Robert Wilson detected faint noise in a radio receiver. Further study showed that these radio waves arrive at earth from all directions, so the radiation must be coming from the universe itself. And the radiation has a thermal signature, showing that it was emitted by something hot. It is the light and heat of the Big Bang. Because of the expansion, the radiation has cooled over time to near absolute zero, consistent with predictions made before it was discovered.

How old is the universe?

Astronomers are able to measure age using several methods. Here are two of them:

Astronomers can calculate how long it will take a star to burn out, based on its size. Big stars burn faster than small stars. A "globular cluster" is a cluster of stars (of different sizes) that formed at the same time. Since the big stars die out first, when only small stars are left scientists can tell that it is an old cluster. So by looking at the size of stars still in the cluster, astronomers can measure its age. The oldest globular clusters found are at least 11 billion years old. The universe as a whole must, therefore, be older than this.

With the Hubble Telescope astronomers can see light that has travelled for about 13.3 billion years, from the very first stars. The universe must be older than this for the light to reach us today.

For further information read: **http://map.gsfc.nasa.gov/universe**

Briefing Sheet Session 1:
Fine-tuning

" The laws of nature have certain constants, and it's not clear why those constants have the values that they do. But it is clear that you can **change those constants a little bit**, and you would have **a universe that's no longer fertile for life**, you'd basically have a sterile universe.

Ard Louis, *Test of Faith* Part 1

" **Our universe** is very particular, **very special** in its character. There's a sense in which the universe was pregnant with life, essentially from the Big Bang onwards, because the very … physical fabric of the world, the laws of nature that science assumes … had to take **a very precise, very finely tuned form** for carbon-based life to be possible.

John Polkinghorne, *Test of Faith* Part 1

" The really important thing is that **the world as we observe it corresponds with what Christians would say the world ought to be like** … there's a correspondence between the theory and the observation.

Alister McGrath, *Test of Faith* Part 1

Fine-tuning, or the Anthropic Principle, is the idea that the universe is fined-tuned for life. There are many different factors that have to be exactly right – otherwise we would not be here. These details have amazed scientists of all religions and none because there is currently no good scientific explanation for why they should all be "set" at such precise values. Here are just a few examples:

1. **Carbon** is an essential element for life. The **strong nuclear force** holds the particles that make carbon together. If the strong nuclear force were any weaker, carbon would never form. If it were any stronger, all the carbon would turn into oxygen. As it is, this balance is tuned exactly so that both elements are present.

2. The number of dimensions in our universe is right for life. You can only have planets with stable orbits if you have three dimensions in space. Any more than three and things would become very unstable, and we could not survive.

3. The amount of **matter** and **energy** present at the time of the Big Bang had to be very finely balanced. If this balance had not been exactly right, the universe would either have collapsed as soon as it began because of the strength of gravity or it would have blown apart too quickly. The amount of matter and energy present had to be correct to an accuracy of 1 in 10^{60} (one with sixty zeros after it).

4. In the universe, **disorder** always increases. The universe must have been much more ordered when it began in order for it to be as organized as it is now. Roger Penrose, a former professor of mathematics at Oxford, calculated that the chance that our universe would have this amount of order randomly is one in ten to the power of 10^{123}. This number is so large that if you were to write a zero on every atom in the visible universe, you would run out of atoms before you ran out of zeros.

5. The **cosmological constant**, often called "**dark energy**," acts as kind of anti-gravity force, pulling the universe apart. It has to have a very small value, very close to what is observed. If it were much greater than it is, the universe would fly apart so rapidly that no stars or planets could form.

6. Atoms are made up of **protons** and **electrons**. The mass of a proton must be almost exactly 1840 times the mass of an electron in order for the building blocks of life, such as DNA, to exist and be stable.

Briefing Sheet Session 1:
Scientists and Faith through History

Thomas Huxley

> " What Thomas Huxley saw in Darwinism was the potential to make biology scientific, to give it a significant public profile and to make science ... important in society in a way in which it wasn't. Part of that meant, in a sense, prizing science from the hands of the clergy who, for various reasons, had tended to dominate scientific positions and the scientific establishment for quite a long time. The strategy he [Huxley] used, you might call it a kind of wedge strategy, was to say, 'actually science and religion are different activities, they're in conflict. Clergy, hands off the science, leave it to the true professionals.'
>
> Peter Harrison, *Test of Faith* Part 1

Thomas Huxley was a British biologist in the 1800s. He was a friend and admirer of Charles Darwin and was a key spokesperson for Darwin's theory of evolution by natural selection.

Many of the people doing science in the 1800s were church ministers. Huxley and eight of his friends formed a dinner group and called themselves the "X-Club." The purpose of the group was to promote science and see it established as a profession supported by public funding. The members of the X-Club were mostly "philosophical naturalists" – people who did not particularly believe in God, but who looked for inspiration in nature. In fact, Huxley came up with the word "agnostic" (someone who does not think there is enough proof about God to decide whether God is real or not).

Huxley and the X-club promoted the idea that Christianity and science had always been in conflict because they wanted to establish a new professional scientific community free of clerical influence. Several best-selling books, which pushed the idea of "warfare" between science and religion, popularized this account. Despite three hundred years of co-operation between religion and science, they created a myth of conflict that people still believe. Today, some scientists still use this story to argue against religion.

> " Perhaps science wouldn't have emerged in the West at all, had it not been for a certain set of religious convictions about how the world was ... the very idea that the world is a place that is rationally intelligible springs from – or at least for these people* it sprang from – the idea that there was a God who had put this order in place.
>
> Peter Harrison, *Test of Faith* Part 1

*(*The key figures in the development of science, most of whom had significant religious commitments.)*

Scientists of faith

Roger Bacon (c. 1214–94)

Roger Bacon was a Franciscan monk who was influential right at the beginning of the development of modern science. He believed it was very important to have an empirical (observed or based on experiment) basis for beliefs about the natural world. He contributed to the idea of "laws of nature." He studied mathematics, optics, the making of gunpowder, astronomy, and the anatomy of the eye and brain.

Johannes Kepler (1571–1630)

Johannes Kepler was an astronomer who formulated the laws of planetary motion that were based on the observations of Tycho Brahe. These are still used to calculate the approximate position of artificial satellites, the outer planets and smaller asteroids. He also did a lot of work in the field of optics and invented a new type of telescope which was used to confirm the discoveries of Galileo.

Galileo Galilee (1564–1642)

Galileo Galilee was one of the early supporters of a sun-centered (heliocentric) view of the solar system. He was censured and imprisoned by the church, but this was mostly because of the way he spoke to people in power. His imprisonment was house arrest, and he was never tortured (as Huxley would have had us believe). He never abandoned his faith and contributed to many areas of science including our understanding of the physics of motion and sound.

Michael Faraday (1791–1867)

Michael Faraday was a chemist and physicist and also an elder in his church. He established the basis for the electromagnetic field concept, electromagnetic induction, and established that electromagnetism could affect rays of light. He discovered benzene and invented the first working electric motors. Some people think he was the greatest experimenter in the history of science.

James Clerk Maxwell (1831–79)

James Clerk Maxwell was a physicist who formulated classical electromagnetic theory in "Maxwell's equations," which synthesized all of the previously unrelated work regarding electricity, magnetism and light into one coherent theory. He demonstrated that electricity and magnetism travel in waves at the speed of light. He also created a statistical way to understand the kinetic motion of gases and laid the foundation for special relativity. Many scientists think that he was as important as Einstein and Newton.

Gregor Mendel (1822–84)

Gregor Mendel was an Augustinian priest and is known as the "Father of Genetics." He studied inherited traits in pea plants and discovered that inheritance follows certain laws. His work went largely unappreciated until the turn of the twentieth century.

Briefing Sheet Session 2:
An Accident in the Making?

CHAPTER 1

Some evolutionary biologists say that the world is without design or purpose. They think that it came into being through a meaningless process, ruled by random chance.

The Bible says that we are made in God's image.

ARE THE BIBLE AND SCIENCE OPPOSED TO EACH OTHER?

Some people are skeptical of anything that predicts a very old age for the earth, and of evolutionary theory.

And some say that living things could not have evolved without any intervention by an intelligent being. They claim that the world was created by an "Intelligent Designer."

CHAPTER 2

But others say that evolution doesn't have to lead to atheism.

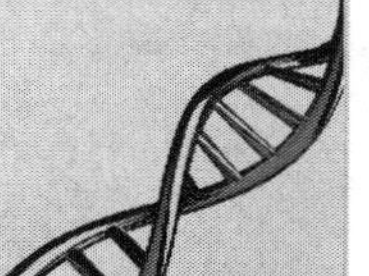

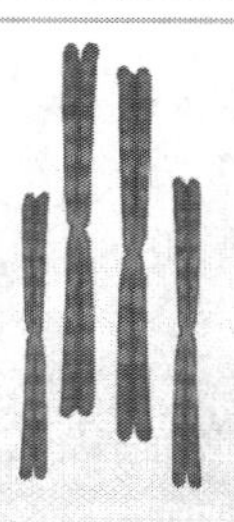

They say that Genesis was meant to be interpreted as an important but not scientific message.

And that you can see reliable evidence for common ancestors in our DNA.

CHAPTER 3

Some people say that evolution is a process **totally up to chance** – like the **random** throw of a dice – at odds with a purposeful God.

But random can mean two things:

1. In day-to-day life we use it to mean "purposeless."

2. In a scientific sense it means that the microscopic details of a process may be unpredictable – but **the overall process may be very predictable**.

So although evolution may appear random, it may be the best way of finding solutions to biological "problems."

In fact, the Professor of Paleobiology Simon Conway Morris believes that evolution can only go in a very few directions – and if you started the process again from scratch, you would end up with very similar things – which fits with the idea that **we were meant to be here**.

CHAPTER 4

But evolution has other challenges for faith – **what about the suffering and death that are part of the process?**

This is the toughest question for Christians in this area.

Some think that the process of producing fruitful life through evolution is a fitting way for God – who loves and gives his people freedom – to create.

Is disease a necessary product of a creation that is able to have life in such variety?

What we *do* know is the good news of the New Testament …

… and that science cannot make paradise on earth. We know that it can be misused, because human beings are flawed creatures.

CHAPTER 5

THE COMMAND GIVEN IN GENESIS WAS NOT TO FIGURE OUT EXACTLY HOW THE WORLD WAS CREATED, BUT TO LOOK AFTER IT.

What we know about climate change must move us to action.

People in the West have benefited from cheap energy in the past. They have a moral duty to reduce their own consumption and help developing countries to develop in sustainable ways.

That change must start with the human heart.

Briefing Sheet Session 2:
Views on Genesis 1

GOD IS CREATOR

1

We should read Genesis 1 as a historical and scientific, common-sense statement of the facts.

The six days in Genesis are twenty-four hours long, so in total God created the world in 144 hours, about 10,000 years ago.

This is the only way to take the Bible seriously. The Sabbath commandment in Exodus that refers to the creation week, and the genealogy of Jesus in Luke, support this view.

Advocates of this view look for scientific evidence that the earth is much younger than mainstream science claims, and that evolution cannot have happened.

This view is incompatible with modern mainstream science and says that mainstream science has interpreted the evidence wrongly because of false assumptions about the physical laws (i.e., that they are always the same through time and space).

For example, there is the idea that small changes may have taken place in animal populations (microevolution) but new species could never form, and that gaps in the fossil record back this up.

ASSUME MIRACLES IN CREATION

2

The "days" of Genesis 1 refer to long periods of time. The Hebrew word *yom* has as many different meanings as "day" does in English. Hebrew does not have a word for a long period of time (era, epoch, etc.), so yom was used instead.

The biblical support for this view comes from the seventh day of God's activity, which is never said to end. This is used by Jesus to clarify the Sabbath law and as a theological theme about heaven by the author of Hebrews. In addition, Scripture teaches that "for God a day is like a thousand years," showing that God measures time differently than we do.

In this view, the events of natural history happened in the order given in Genesis 1, but were stretched out over much longer periods of time. This is consistent with the billions-of-years time frame of mainstream science, but the order of events is somewhat different. God miraculously intervened at some points during the development of creation, such as the creation of plants or birds. (These were not created in the order suggested by evolutionary biology.)

VERY OLD UNIVERSE, LONG TIMESCALE OF CREATION

3

There are different types of literature in the Bible: history, songs, poems, parables, etc. Genesis 1 should be interpreted with an eye for literary devices such as repetition and figurative language, and with an understanding of cultural, historical and biblical context.

For example, the sun and moon are not called by their proper names, because these names also referred to gods in the surrounding pagan cultures. Instead, they are called "big lamp" and "small lamp" to emphasize that there is only one God. The narrative is structured around God creating spaces by separating things, then filling those spaces:

SEPARATION	FILLING
Day 1, Light and darkness	Day 4, Sun and moon
Day 2, Sky and sea	Day 5, Birds and fish
Day 3, Sea and dry land	Day 6, Animals and humans

In this view, Genesis is not a scientific text. We should look first at what the text meant to the first audience to learn its non-scientific message (the "who" and "why"), then at modern science to understand how and when God created the universe, the earth and life.

Briefing Sheet Session 2:
Is there Purpose in Evolution?

Converge

1. To come together from different directions so as eventually to meet.
2. Converge on/upon: to come from different directions and meet at.

Stephen Jay Gould was a world-class palaeontologist and writer of some very popular science books. He didn't believe in God, and he thought that evolution was a purposeless, undirected process. He thought that if you "re-ran the tape of life," and let evolution happen all over again, you would end up with something very different:

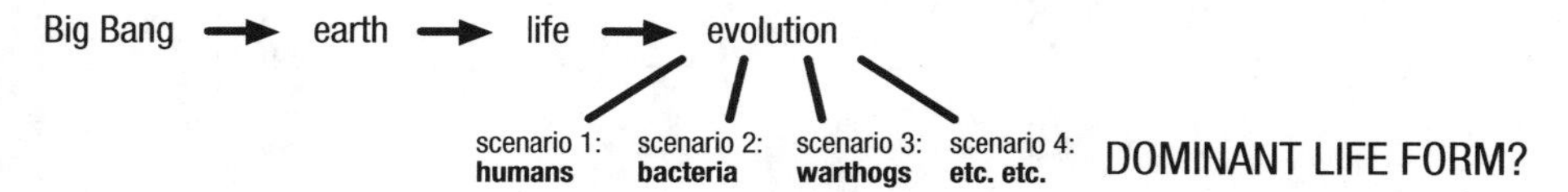

Simon Conway Morris worked on the same type of fossils as Stephen Jay Gould (from the Burgess Shale in British Columbia, Canada) and came to a very different conclusion. **He thinks that there are only so many ways that the process of evolution (on any planet) could "make" things.** The evidence he points to is ***"convergence."***

For example, there were very similar sorts of animals in North America and Australia (some now extinct) – with one important difference. The North American versions were placental mammals, and the Australians were marsupials (with pouches like kangaroos). So there were marsupial moles, flying squirrels, wolves, mice and, in South America, marsupial saber-toothed cats.

What does this show? These two types of animals evolved independently but acquired very similar characteristics. So the process of evolution is not completely random but converges on certain solutions to a problem (of how to "make" effective diggers, treetop dwellers, or predators).

It could look as if things were meant to be that way ...? Is it on purpose?

> “ The general received wisdom is that **evolution is completely open-ended. It can go in any direction you like**. And at first sight that seems entirely reasonable, because biology – evolution – seems to be largely without any underlying laws. So leading from that there is then a metaphysical assumption which says that all things are equally probable, but also everything is accidental: it could be this, it could be that. Whereas **my view is the exact reverse, that the roads of evolution are well defined, and evolution can go in only a very few directions** in comparison to the immense plenitude which the neo-Darwinian would suggest.
>
> “ Effectively it is arguing that from different starting positions, some sort of structure emerges which is remarkably similar. In other words, I go to an aquarium, I see an octopus in the tank. I look at the octopus, and she glances back at me. The eye in my head is constructed on what we call a camera principle. The eye in the octopus' head is constructed effectively the same way; it's also a camera eye. Now we know enough from molecular biology, from the fossil record, and evolution generally, that the common ancestor – and indeed we do have a common ancestor with the octopus – lived about 500 million years ago in the Cambrian, and could not possibly have possessed that sort of camera eye. So **independently the ancestors of the octopus, and the ancestors of Simon Conway Morris, generation by generation, evolved a system which is extremely effective for seeing, called the "camera eye."** So that's just one example of convergence. It goes a bit further than that because this camera eye has actually evolved independently, I think, about seven times, so you almost get the sense that there are stable points in biological space towards which things can navigate.
>
> Professor Simon Conway Morris, *Test of Faith* Part 2

Further Reading

Simon Conway Morris, "Extraterrestrials: Aliens like Us?"
http://adsabs.harvard.edu/abs/2005A&G....46d..24M

Simon Conway Morris (ed.), *The Deep Structure of Biology: Is Convergence Sufficiently Ubiquitous to Give a Directional Signal?* (Templeton Foundation Press, 2008)

Simon Conway Morris, *Life's Solution: Inevitable Humans in a Lonely Universe* (Cambridge University Press, 2003)

Simon Conway Morris, *The Crucible of Creation* (Oxford University Press, 1998)

Briefing Sheet Session 2:
Views on Genesis 2 and 3

This sheet outlines the different ways that the biblical account of the creation of humankind and the scientific account of human evolution might relate to each other. Christians may well hold views that combine elements of several of these. The Adam and Eve narrative sets the scene for the fall, so view A on the fall goes with view A on Adam and Eve, and so on.

Genesis 2: Who were Adam and Eve?

A. We should read Genesis 1 as a historical and scientific, common-sense, statement of the facts. God created Adam and Eve miraculously on the sixth day of creation.

B. We should read Genesis 1 as a historical and scientific, common-sense, statement of the facts. God intervened in a miraculous way at several points in evolutionary history, including at the creation of two human beings: Adam and Eve.

C. While the early chapters of Genesis are not a historical document in the modern sense, they do refer to events that really happened. But they happened in the culture and place Genesis describes. God chose a couple of Neolithic farmers (Adam and Eve) in the ancient Near East (or maybe a community of farmers) and revealed himself to them in a special way, bringing them into fellowship with himself. They were representative of all humankind, as the first people God brought into relationship with himself.

D. As in Model C, while the early chapters of Genesis are not a historical document, they do refer to events that really happened. Among early humans, there was a growing awareness of God's presence and calling upon their lives to which they responded in obedience and worship.

E. Genesis 1 should be interpreted with an eye for literary devices such as repetition and figurative language. There is no historical connection between the theological and biological stories. The question of the birth of the first spiritually alive humans is essentially unanswerable. The purpose of the early chapters of Genesis is to give a theological account of the role and importance of humankind in God's purposes.

Genesis 3: Views on the fall

A. There was no death at all before the fall. When Adam and Eve disobeyed God not only did they die spiritually, but there were also big changes in the way creation operates. From that point on, decay and the physical deaths of animals, plants and humans were possible.

B. Until the fall, Adam and Eve were immortal. When they disobeyed God they died spiritually, and later physically. The physical deaths of animals and plants were already occurring.

C. Adam and Eve (or the group of people God chose) disobeyed him. This act of disobedience separated them from God and they died spiritually. Because these people were representative of all humankind, everyone else fell too. The physical deaths of humans happened throughout evolutionary history.

D. The fall was a conscious rejection of the growing awareness of God's calling. It led to spiritual death. The physical deaths of humans happened throughout evolutionary history.

E. This is the eternal story of us all. It is a theological account that describes the common experience of separation from God through disobedience to God's commands. The result for us is spiritual death. The physical deaths of humans happened throughout evolutionary history.

Briefing Sheet Session 2:
The Science behind Climate Change

(Based, with permission, on the JRI briefing paper by Sir John Houghton, "Global Warming, Climate Change and Sustainability: Challenge to Scientists, Policy-makers and Christians" [2007] **www.jri.org.uk**)

"Greenhouse gases" in the earth's atmosphere such as water vapor, carbon dioxide or methane, trap heat and keep the earth warm. This "greenhouse effect" keeps the earth 20 – 30°C warmer than it would otherwise be and is essential for our survival. But the greenhouse effect is increasing. We have a record of what the weather was like in the past, and of the gases in the air at that time, preserved for us in the ice caps in Greenland and the Antarctic. Scientists can drill down through the layers of ice that have built up over thousands of years and analyze the bubbles of gas trapped in each layer. From this we can see that, since the beginning of the industrial revolution in the 1750s, the amount of carbon dioxide in the atmosphere has increased by nearly 40%. With chemical analysis we can see that this is mostly because of the burning of fossil fuels (coal, oil and gas).

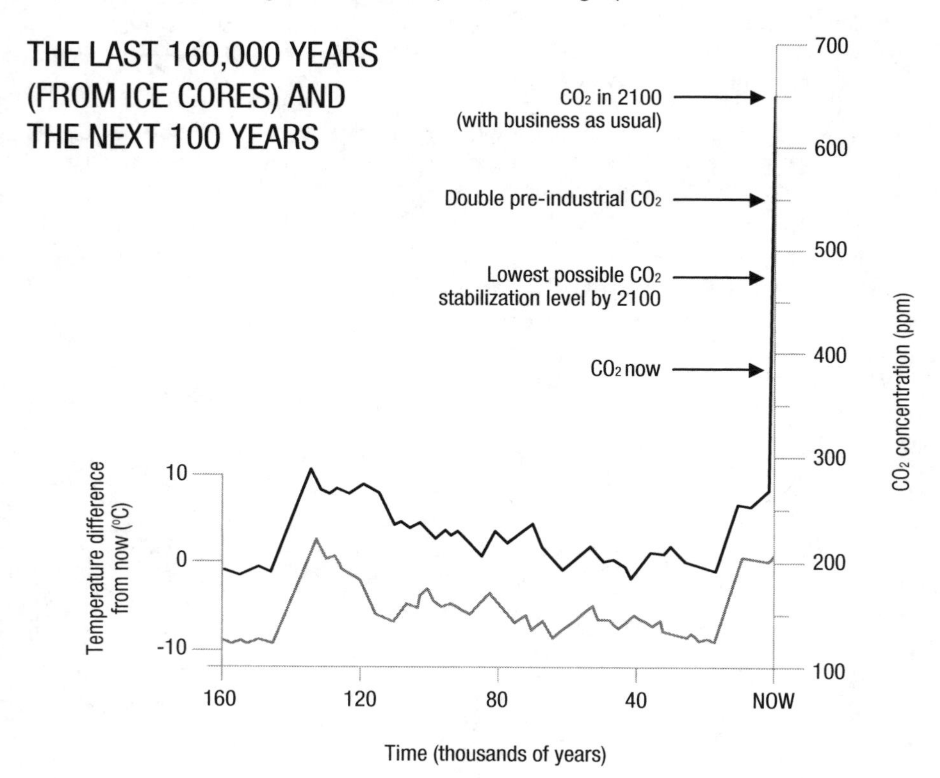

The average temperature on earth has risen over the last century. There is strong evidence that most of this rise has been caused by the increase in greenhouse gases, and especially carbon dioxide. Scientists predict that, during the twenty-first century, the average temperature will rise by 2-6°C. This doesn't sound like very much, but the difference in average temperature between the middle of an ice age and a warm period is only about 5-6°C. The predicted temperature rise could have a huge impact.

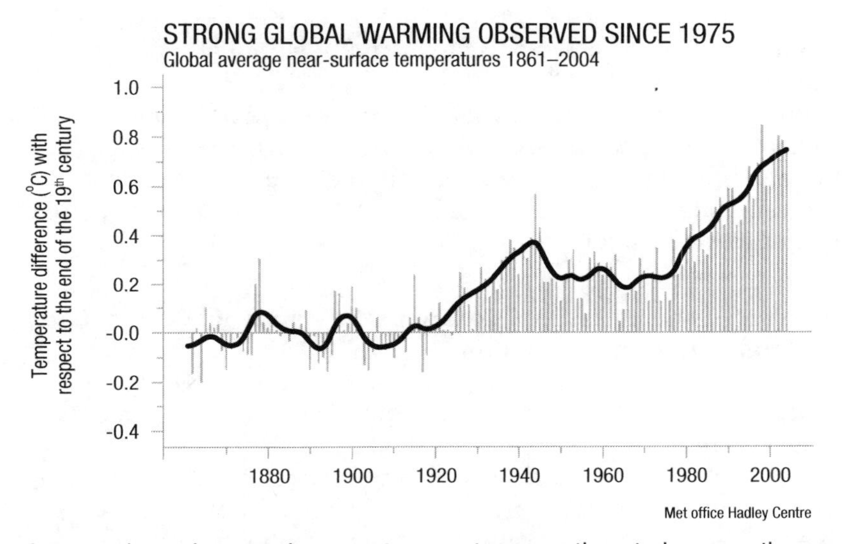

As water heats it expands, so as temperatures continue to increase the sea level will rise, flooding low-lying coastal areas around the world. The temperature changes already produced by humans will take hundreds of years to feed into the deep ocean, so the sea level will continue to rise for hundreds of years even if we stopped producing any more greenhouse gases overnight. Warmer temperatures will also cause greater evaporation of fresh water on land, leading to more water vapor in the atmosphere and more rain or snow. This will cause drought in some areas and flooding in others. There is no evidence that hurricanes will become more common, but it is possible that they will become more severe as the surface temperature of the sea increases.

All of these changes in the weather will affect the ability of humans, plants and animals to survive. The worst impact will be felt in developing countries. In the short term, crop yields will increase in colder countries, but the damaging effects in warm countries, flooding and storms will far outweigh these advantages. Eventually crop yields will decrease worldwide as temperatures increase further. If we cut down our production of greenhouse gases now, the harmful effects will be greatly reduced. It has been argued that developed countries, which have benefited from burning huge amounts of fossil fuels, should make the biggest efforts to cut down and allow developing countries to continue to develop.

Briefing Sheet Session 2:
Climate Change Questions

(Adapted with permission from "Climate Change Controversies: A Simple Guide," The Royal Society [2007] **http://royalsociety.org**)

Q1: The earth's climate always varies. Aren't we just in a natural period of warming?

A: The earth's climate varies due to many different factors, including cycles of ice ages caused by changes in the distance between the earth and the sun, volcanic eruptions and changes in the sun itself. However, none of these factors is enough to explain the rapid changes in the last 100 years.

Q2: There isn't enough carbon dioxide in the atmosphere to cause any significant change, is there?

A: Although there isn't a big volume of carbon dioxide (CO_2) in the atmosphere, it can have a significant effect. It has a direct effect because it traps heat very strongly. It also has an indirect effect because, as the earth warms up, water evaporates more quickly from lakes and the sea. This increases the amount of water vapor in the atmosphere, which causes an even stronger greenhouse effect.

Q3: Isn't the increase in carbon dioxide in the atmosphere the result of climate change, rather than the cause?

A: As the oceans and soil warm up they do release carbon dioxide into the atmosphere. Scientists can find the origin of carbon dioxide in the atmosphere through chemical analysis. Most of the increase in CO_2 levels comes from burning fossil fuels.

Q4: I thought that the observations of weather balloons and satellites were inaccurate?

A: In the early 1990s there were errors both in the way that data was collected and in the way it was analyzed. These errors have been corrected, and now the data from weather balloons and satellites agrees with data collected by other methods.

Q5: Aren't computer models of the climate inaccurate?

A: Although the climate is very complex, scientists have been able to create increasingly accurate models of the way it works. These computer models have been used to simulate changes in the climate over the course of the last century, and their simulations have matched what actually happened. Using these models scientists can give general predictions about the course of the climate in the future on a global scale, based on different predictions about human behavior.

Q6: Isn't climate change caused by the sun becoming more active?

A: The sun's activity does play a role in shaping climate. However, that alone is not enough to explain the recent warming. Also, there has been very little change in the sun's activity over the last three decades, so this cannot account for the observed warming.

Q7: Surely it's not a big deal. Aren't climate scientists exaggerating?

A: The earth's ecosystems are very finely balanced. Even a change of 2-3°C would be greater than has been seen for ten thousand years, and many species would find it very difficult to adapt. The people most affected will be those in developing countries and the poor, creating greater inequalities in access to food, clean water and medical treatment.

BRIEFING SHHETS

Briefing Sheet Session 3:
Is There Anybody There?

CHAPTER 1

Some scientists claim that we run on rails determined by our physical characteristics, and that spiritual experiences are just a by-product of our brains.

But neuroscientists who are also Christians believe that we are made in the image of God.

When we are having a religious experience, something happens in our brains.

But that doesn't prove that religious experience is **just** what is in our brain. The experience also means something.

And it is possible to artificially stimulate certain parts of the brain, and make someone feel angry or sad.

But that doesn't mean that it isn't possible to feel happy or sad for genuine reasons!

CHAPTER 2

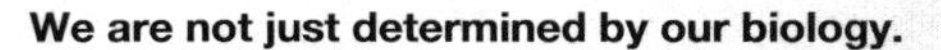

We are not just determined by our biology.

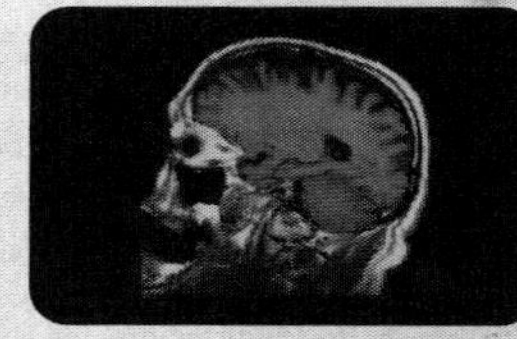

Our experiences and the environment we live in also play a role in shaping who we are.

And these scientists believe that the brain is more than the sum of its parts.

You could say that **our minds and personalities emerge from the complex structure of the brain**.

But they can't be defined by just looking at the cells in the brain.

It's like music that emerges from the strings of an instrument – it can be described in a mechanical way, but it has a much deeper meaning than vibrations of strings.

CHAPTER 3

With genetic technology, we now have the ability to enhance or even clone ourselves.

But **even a genetic clone would be more than the sum of its DNA**.

We are creative	We have the ability for moral reasoning	We make meaningful choices

These abilities seem to go far beyond our biological make-up.

CHAPTER 4

To really understand things in a meaningful way, theology and neuroscience need to talk to each other.

We need to read both the book of the Bible and the book of nature and not be afraid of asking new questions.

We need to use the tools of **both** faith and science to explore the world.

And, **for Christians, science increases their sense of wonder at what God has created**.

BRIEFING SHHETS

Briefing Sheet Session 3:
When Does Human Life Begin?

Below is a summary of the main views held on the status of the early embryo and the arguments that people use to defend them.

A) Human life begins at fertilization *(0–6 hours)*

Biblical/theological arguments

- The Bible names Jesus and other people by this stage.
- This is the origin of a "personal history."
- This is when Jesus became incarnate as a man.
- Relationship with God is established.
- In the Bible the Hebrews believed life began as soon as they were aware of it being there. The message is that life begins as soon as there is something there. With our knowledge today, this means conception.

Biological arguments

- Fertilization provides a fairly precise moment of beginning.
- The genetic make-up of the individual is specified during this stage.

B) Human life begins at implantation *(7–10 days)*

Biblical/theological arguments

- Physical relationship with the mother begins – she can become aware of her pregnancy. Part of what defines us as human is being in relationship.
- Twinning may occur between the "blastocyst" (hollow ball of cells) stage and implantation, so until implantation there isn't "one" individual present to relate to God.
- This is what passages about the unborn refer to because this is the time when pre-scientific societies became aware of pregnancy.

Biological arguments

- There is a high rate of embryo loss before implantation (70–80%).
- Until implantation, it is impossible to tell what parts will become the embryo and what will become the placenta.

C) Human life begins at the primitive streak stage *(14 days)*

Biological argument

- The development of the primitive streak marks where the nervous system will begin to develop. The capacity for sensation and pain are important in defining humanness and in determining how we treat others.

D) The beginning of human life is a continual process

This is the view that all of the above "milestones" are not that critical, since the development of human life is a continual process from fertilization through to birth and onwards. Human life deserves our care and protection all the way through, although prenatal care will increase in line with development.

Some Bible passages that are relevant to this discussion

These highlight the fact that human development is shaped and purposed by God from the beginning:

- Jesus' incarnation – Luke 1:31–33
- God establishes a relationship with Isaiah and Jeremiah before birth – Isaiah 49:5; Jeremiah 1:5
- God's knowledge of us in the womb – Psalm 139:13–16 and Job 10:8–12
- An important marker of new life was "quickening," when a baby kicked for the first time – Luke 1:44
- Being in relationship is an important part of personhood – Genesis 2:18
- Care of pregnant women – Exodus 21:22–23

Endnotes

i. Figures from **http://esa.un.org/unpp**.

ii. **www.abc.net.au/rn/allinthemind/stories/2006/1698423.htm**.

iii. Examples and scenario taken with permission from a paper by D. Alexander called "Cloning Humans: Distorting the Image of God?" (The Cambridge Papers 10.2 [June 2001]; **www.jubilee-centre.org**, Jubilee House, 3 Hooper Street, Cambridge, UK, CB1 2NZ).

iv. Extract with permission from D. Alexander and R. S. White, *Beyond Belief: Science, Faith and Ethical Challenges* (Lion, 2004), pp. 152–3.

v. Copied, with permission, from *Life in Our Hands: A Christian Perspective on Genetics and Cloning*, by John Bryant and John Searle (IVP, 2004) p. 152.

vi. "GM" in this case refers to "Somatic" gene therapy – a treatment that affects a tissue or tissues in the body, but not the eggs or sperm, so that the treatment does not affect the next generation (because that raises a host of other issues and is illegal in the UK).

vii. See note above.